The Chemical Story of Olive Oil
From Grove to Table

The Chemical Story of Olive Oil
From Grove to Table

Richard Blatchly
Keene State College, New Hampshire, USA
Email: rblatchly@keene.edu

Zeynep Delen Nircan
Boğaziçi University, Istanbul, Turkey
Email: zeynep@egedeatolye.org

Patricia O'Hara
Amherst College, Massachusetts, USA
Email: pbohara@amherst.edu

THE QUEEN'S AWARDS
FOR ENTERPRISE:
INTERNATIONAL TRADE
2013

Print ISBN: 978-1-78262-856-9

A catalogue record for this book is available from the British Library

The Royal Society of Chemistry is a charity, registered in England and Wales, Number 207890, and a company incorporated in England by Royal Charter (Registered No. RC000524), registered office: Burlington House, Piccadilly, London W1J 0BA, UK, Telephone: +44 (0) 207 4378 6556.

Visit our website at www.rsc.org/books

Printed in the United Kingdom by CPI Group (UK) Ltd, Croydon, CR0 4YY, UK

Preface: Our Olive Origin Story

I am never satiated with rambling through the fields and farms, examining the culture and cultivators, with a degree of curiosity which makes some take me to be a fool, and others to be much wiser than I am.

April 11, 1787, Nice, France – letter from Thomas Jefferson to his friend and French military officer, the Marquis de Lafayette.

Our last few years have been spent "rambling through the fields and farms" in our quest to truly understand the link between olive cultivation, olive oil production, and olive chemistry. Our journey began in 2010 with a summer workshop in Turkey exploring "Zeytin" (Turkish for "The Olive").

ZDN: I was so excited and inspired by the idea of a liberal arts approach to education, which I understood to be holistic and interdisciplinary learning with praxis that must be in perfect harmony with the cultural character of learners in a community. I reasoned this could lead to a better future in a better society. I felt compelled to find out if now, or at any point in history, such an approach existed in my country (I felt certain it had to) and to build a community of scholars (not just chemists) who could create a modern version of it. With this idea, I started to contact scholars of the highest caliber from around the world. The natural first step was to bring some of them together and create an example. I chose Yeni Foça, which I considered to be my

The Chemical Story of Olive Oil: From Grove to Table
By Richard Blatchly, Zeynep Delen Nircan and Patricia O'Hara
© Richard Blatchly, Zeynep Delen Nircan and Patricia O'Hara, 2017
Published by the Royal Society of Chemistry, www.rsc.org

hometown, as the pilot workshop location. Of course, I anticipated the invisible cultural barriers that would inhibit conversation among people from diverse backgrounds. What could be a
topic that is near and dear to almost everyone on earth? It turns
out the perfect answer to this question is ... The Olive! I was exhil-
arated to see how much enthusiasm this idea brought to everyone
and the community of people it brought together. I named the
endeavor Ege'de Atölye, for which O'Hara and Blatchly (among a
few others) became advisors. The two, whom I admired, were naturally the first to be invited to the upcoming "Zeytin" workshop
to create the olive centered chemistry module, which then grew
into a journal article, (R. A. Blatchly, Z. Delen and P. B. O'Hara,
Making Sense Of Olive Oil: Simple Experiments to Link Sensory
Observations with Underlying Chemistry, *J. Chem. Ed.*, 2014, DOI:
10.1021/ed300557r), courses taught at universities in Turkey and
in the USA, two blogs (www.worldolivepress.blogspot.com and
www.egedeatelye.blogspot.com), almost two dozen public talks,
and now this book. Since then, Ege'de Atölye, too, continued to
grow with more workshops, friends, and many other interesting
projects. (New workshop invitations and a catalog of old ones are
posted at www.egedeatolye.org) Despite the financial and professional sacrifices, it became the best occupation I could ever
imagine, allowing me to maintain the perfect balance between
independency from and belonging to academia.

From the perspective of Turkey, I think this book partially ful-
fills the original dream and therefore makes me incredibly proud.
First, it is proof that scholarly learning is possible anywhere and
can be fun without expensive facilities and highly paid faculty. It
is possible to find wonderful faculty (explorers like POH and RB)
who would be willing to participate voluntarily. Second, this is not
just a book about olives or chemistry but hopefully it establishes
a deeper connection between Turkey and the world by giving a
glimpse of a Turkey different than what one usually sees in international media. Finally, nothing would make me happier if this
book intrigued or inspired a few young Turks to innovate around
the olive and other precious native products towards a brighter,
self-sufficient future, both for themselves and the country.

POH and RAB: The chance to visit our friend and colleague in
Turkey, and to both learn about and perhaps contribute to their
educational system, was too good to pass up. As we explored the

chemistry of olive oil, we were able to move past our initial impression of simplicity (it is, after all, mostly a rather simple mixture of triglycerides) and discover the unique chemical qualities that allow it to stand alone among edible oils. We were fortunate to be starting as the chemical story was coalescing in labs around the world, the world's consumption was exploding, and the stories about fraud were making headlines (such as Tom Mueller's book *Extra Virginity*). As we learned from producers in Turkey about the standards for extra virgin olive oil, about the bitterness and pungency of the best oils, about how the oil is produced, about the fraud, and about the amazing connection between human culture and the olive, we were inspired to learn more. Our background as educators inspired us to share the story with as many as we could. Since then, we have devoted significant time, including a year on sabbatical, to compile research, and to talk with cultivators, producers, and scientists. We could not have done this work without the extraordinary gifts of time we received from so many around the world.

ALL: The research world continues to (re)discover the health benefits of extra virgin olive oil, and it seems a week doesn't go by without some new health claim or allegation of fraud. As we traveled and spoke to those not in the industry, people wanted to know what they could believe. We became convinced that the answer was to convey what makes olive oil so special from a chemical perspective, and then make the connection to everyday use. We dedicated ourselves to reading through the many claims about health or the uses of olive oil and translating them into plainer language. The more we did this, the more grateful our audiences and the more we were urged to write this up to share with others. At the same time, the farmers and millers we met were eager to share their knowledge and passion about the olive and curious to understand the links between quality extra virgin olive oil and chemistry.

The decision to write this book was made in yet another rental car on our way to yet another olive growing town – this one, Kritsa, on the island of Crete in 2014. We realized that along the way something had happened to us that is true of many people who begin an inquiry into olives, olive oil, or olive trees. We had become true disciples of the olive that holds so many truths under its spreading branches and unique fruit.

Nazim Hikmet 1902–1967

ON LIVING	YAŞAMAYA DAİR
Living is no laughing matter: you must live with great seriousness like a squirrel, for example—I mean without looking for something beyond and above living, I mean living must be your whole occupation	Yaşamak şakaya gelmez, büyük bir ciddiyetle yaşayacaksın bir sincap gibi mesela, yani, yaşamanın dışında ve ötesinde hiçbir şey beklemeden, yani bütün işin gücün yaşamak olacak
Living is no laughing matter: you must take it seriously, so much so and to such a degree that, for example, your hands tied behind your back, your back to the wall, or else in a laboratory in your white coat and safety glasses, you can die for people—even for people whose faces you've never seen, even though you know living is the most real, the most beautiful thing	Yaşamayı ciddiye alacaksın, yani o derecede, öylesine ki, mesela, kolların bağlı arkadan, sırtın duvarda, yahut kocaman gözlüklerin, beyaz gömleğinle bir laboratuvarda insanlar için ölebileceksin, hem de yüzünü bile görmediğin insanlar için, hem de hiç kimse seni buna zorlamamışken, hem de en güzel en gerçek şeyin yaşamak olduğunu bildiğin halde
I mean, you must take living so seriously that even at seventy, for example, you'll plant olive trees—and not for your children, either, but because although you fear death you don't believe it, because living, I mean, weighs heavier	Yani, öylesine ciddiye alacaksın ki yaşamayı, yetmişinde bile, mesela, zeytin dikeceksin, hem de öyle çocuklara falan kalır diye değil, ölmekten korktuğun halde ölüme inanmadığın için, yaşamak yani ağır bastığından

Acknowledgements

This book is the product of a collaborative project. We owe much of what we learned about olives and olive oil to the hundreds of grove owners, producers, scientists, and professionals who gave so generously of their time. We are indebted to you all. We hope we have faithfully told your stories.

We especially wish to thank the following individuals for providing feedback to us for individual chapters: Dimitri Gutas; Professor Arabic and Greek Arabic Studies, Yale University, New Haven, USA; John Chandler, Headmaster *emeritus*, Robert College, Istanbul, Turkey; Xavier Rius, Engineer of Agriculture at Agromillora, Spain; Arend Hofmeyr, owner Portion 36, Stellenbosch, South Africa; Mücahit Taha Özkaya, Professor of Agriculture, Ankara University, Turkey; Cahit Tunç, Retired R&D Chief of Tariş Cooperative, Turkey; Mayo Ryan, Producer, Vice President, California Olive Ranch, USA; Hakan Barçın, Olive Oil Producer, Cofounder of Taşköy, Foça, Turkey; Virginia Brown Keyder, International Law Specialist, Instructor at Boğaziçi, Sabancı and Birmingham Universities, Turkey and USA; Selina Wang, Professor UC Davis, leader of UC Davis Olive Oil Laboratory, Davis, CA, USA; Selin Ertür, Selatin Olive Oil Producer, Oleologist, MIPAAF Official Olive Oil Taster, Turkey; Sue Langstaff, owner Applied Sensory, LLC and Olive Oil Sensory Panel Leader, Davis, CA, USA; Kevser Özden Pişkin, Professor of Biochemistry, Hacettepe University, Ankara Turkey; Linda Costa,

The Chemical Story of Olive Oil: From Grove to Table
By Richard Blatchly, Zeynep Delen Nircan and Patricia O'Hara
© Richard Blatchly, Zeynep Delen Nircan and Patricia O'Hara, 2017
Published by the Royal Society of Chemistry, www.rsc.org

Author, Taster, Stellenbosch, South Africa; Zerrin İren Boynudelik, Author, Professor Art History Yıldız University, Istanbul, Turkey. Their help shaped the work you have read in only positive ways. Thank you! We alone are responsible for any misunderstandings or misrepresentations that remain in the text.

Another source of valuable feedback came from students in our classes taught at Keene State College Fall 2015 HONORS 290 and Boğaziçi University Fall 2014, Summers 2015 and 2016 CHEM 485. We find our students to be a constant source of inspiration. Our favorite student quote after reading an early version of the planting chapter is: "*I had NO idea that there were so many decisions that needed to be made before you could plant a tree!*"

Across the world, we were fortunate to visit five continents and nine countries. We should include each and every one of the several hundred people who gave us so much of their time, but space precludes our doing that. Instead we will highlight a few of the many, and let them stand for all of you.

We spent the majority of our time together in Turkey and we feel the most indebted to those who helped launch this project. Besides those mentioned above, the following people were instrumental to our learning: Artun Ünsal, Ümmühan Tibet, and Murat İsfendiyaroğlu, who enlightened and intrigued us as olive and olive oil experts. Meltem Türköz, who made us see the olive and olive oil from a local and cultural perspective along with Ulrike Muss and İklil Erefe Selçuk. Zerrin İren Boynudelik, who introduced us to the archeological and historical perspectives of olives. Funda Barbaros, Meneviş, and Uzbay Pirili helped us understand the economical perspective of olive oil production. Christian Wernz, Ahmet Kutsi Nircan, Ali Rana Atılgan, and Dora Üretken challenged us to think about the engineering perspective of olive oil production. Derya Ülker and Burcu Delen helped us to interact with olives through the performing and visual arts.

We are especially grateful to the residents of Yeni Foça and staff of Nesin Math Village, who welcomed us during several Ege'de Atölye Zeytin workshops. Esen Çeşmeci helped to initiate and facilitate our work with the Foça Municipality and Mayor Gökhan Demirağ supported and welcomed us all. I (ZDN) am especially indebted to Ege'de Atölye advisors John Chandler and Hülya Denizalp for motivating me to keep organizing the workshops despite all obstacles. These workshops kept our curiosity and motivation alive for many years and sped up our learning immensely, so it is hard to think of this book becoming a reality without them. The workshops were also made possible thanks to Türk Kültür Vakfı and all members and supporters of Ege'de Atölye, some of whom are listed in the www.egedeatolye.org website.

The Chemistry Department at Boğaziçi University in Istanbul is home to ZDN as a part-time instructor and hosted the Fall 2014 sabbatical of two of us (POH and RAB). We are grateful to Viktorya Aviyente, Chair of the Chemistry Department, for making us feel welcome and to Taner Bilgiç, Dean of International Studies and Professor of Industrial Engineering, for helping us with class enrollments, visas, and collaborations across campus. The faculty and staff in the Department of Chemistry were instrumental in making us feel at home.

And, finally, Professor Halil İnalcık, who unfortunately died at age 100 before he could see the publishing of this book, is remembered here for his continuous scholarly and personal encouragement.

While in the Northern Hemisphere in the fall of 2014, we traveled to countries in the Mediterranean to explore harvest and production practices. We know this is by no means a representative sample of groves but cherish the invitations that led us to visit a particular place, as they came with a deeper look at the culture that surrounds the production of oil. We know we have missed whole regions and look forward to a time when we can visit them all.

- In Catalonia, special thanks to Esther Gelabert, Belianes Consulting Spain, for fostering and organizing international collaborations and to Mayor Josep Ramon and the Municipality of Belianes for inviting us to speak at their Fall 2014 Olive Oil Festival. We were happy to meet Xavier Rius from Agromillora, the largest nursery in the world, and tour their incredible nursery and labs outside of Barcelona.
- In Italy, thanks to the American Field Service office in Colle di Val d'Elsa for helping us to arrange our trip and get us started. We visited olive groves from small farm owner Guido Tinacci, whose farm goes back 600 years, to larger groves in Principe Corsini in Tuscany. The Mediterranean fruit fly had

decimated the crop in Tuscany, and it was a sad but realistic picture of what happens when things don't go right.

- In Greece, special thanks to Aris Kefalogiannis of Gaea Olive Oil for helping arrange our visit with Nikos Zachariadis, Director of the Agricultural Cooperative of Kritsa on the island of Crete. Nikos toured us through a most efficiently run olive press at peak harvest time and introduced us to the local cuisine, along with the deft translations from Joanna Sarantopoulou. Our host Argyro Tzanakis made us feel like family during our stay with her, and we've promised her we will return.

Two of us (POH and RAB) continued our sabbatical by taking up visiting professorships at University of Stellenbosch in South Africa. This allowed us to experience a second harvest season in the southern hemisphere. We are deeply grateful to Michael Schmeisser, Professor in the Horticultural Department, for hosting our spring 2015 stay. Sitting in on his lectures and being exposed to faculty and students in the department taught us so much about plants and life in South Africa. Mohammed Karaan, Dean of the Faculty of AgriSciences, took a particular interest in our project. We miss our early morning brainstorming sessions with him about the importance of higher education in fostering leadership and entrepreneurial endeavors through projects such as ours. Jonathan Scrimgeour, owner of Buffet Olives in Paarl, invited us early in our stay to meetings of the South African Olive Association and helped us network with South African growers and producers. Arend and Birgitta Hofmeyr, owners of Portion 36 Olive Grove, shared hours with us, usually accompanied by an amazing meal as we picked their brains about tree and grove management and experienced the best ever olive oil ice cream with olive oil praline cookies and fresh picked Catawba grapes. Linda Costa was a valuable friend and encouraged us to follow our instincts and be ambitious in our explorations of the South African olive country.

Our grove visits in South Africa started with a St Patrick's Day harvest with Bryan Beverley and his team making L'Ormarins oil from sky-high groves at the Antonij Rupert estate. In the late afternoon we went with our picked olives to Loesje Kock's press in nearby La Bourgogne Farm. At Tokara's Olive Shed near Stellenbosch, Robert Claasen and operations manager Gert gave us a full tour and tasting. We visited the Morgenster Estate in Sommerset West, where we met up with owner Giulio Bertrand and saw operations with Chris van Niekerk. We toured the groves with farm manager Corius Visser and did a tasting with Judi Dyer. Rob Still of De Rustica hosted a fantastic stay for us in the Klein Karoo, and operations manager Jup Steenkamp shared with us some of his experiments in malaxing and filtering. Barry Anderson and Adriaan David showed us operations at Gariëlskloof (also a fantastic restaurant), and at nearby Anysbos, Johan Heyns gave us a tour of the olives and cheese making station. Sean and Rene White at The Greenleaf Grove gave us a tour of one of the largest groves and presses in South Africa in Malgas, near Cape Algulhas at the very tip of Africa. Anders Rabie at Willow Creek near the Nuy Valley showed us around a newly expanded modern facility. Nick Wilkinson's Rio Largo Press in the Scherpenheuwel Valley was unique in the way every aspect of the milling process can be monitored and adjusted. Thanks also to Brenda Wilkinson for putting us in touch with Tom Mueller. We also enjoyed getting back into the lab and seeing operations with Raymond Hartley at MicroChem Labs in Capetown.

An invitation from Gerri Nelligan, Editor of the Australian & New Zealand Olivegrower & Processor Newsletter, inspired our quick visit to Australia and New Zealand. In Australia, Paul Miller, former President of Australian Olive Growers Association, helped us with introductions and logistics. Thanks to Claudia Guillaume at Modern Olives lab in Lara for showing us the accredited lab and sensory panel facility; to Jim Rowntree at Tatiara Olive Press in Keith and his wife Lisa, Executive Director of Australian Olive Association, who together with sons and daughter run Longridge Estate in Coonalpyn; to Mike Smith and Leandro Ravetti at Boundary Bend in Murray for allowing us to spend two days with them touring facilities and groves; to Jamie Ayton, Belinda Taylor, and Helen Taylor at the Edible Oil Laboratory in Wagga Wagga; to Peter and Caroline O'Clery at Homeleigh Olives who met us early on a cold Saturday morning at the Canberra Farmer's Market; and finally to Jeff Konstantino from Fedra Olive Grove. All of you deserve our most sincere thanks.

In New Zealand, Andrew Taylor and Gayle Sheridan of Olives New Zealand helped arrange our visits. Thanks to Bruce McCallum and Billy Hey at The Olive Press who were in the middle of moving their press to a new building; to Mark and Kate Bunny from Loopline Olives in Wairarapa; to Wayne Startup and Tom Casey at The Village Press in Hawkes Bay; to Margie and Ross Legh at The Olive Place in Whangarei; to Anne and Collin Stanimiroff at Rangihoua Grove and Press; and finally to Margaret and John Edwards at Matiatia Grove and Press in Waiheke Island.

In the United States, two of us (POH and RAB) enrolled in a course on Sensory Evaluation of Olive Oil at UC Davis in CA. Thanks to Sue Langstaff who through lectures and hands on tasting taught us how to detect and describe sensory defects. Our favorite Sue quote is "Guys, it's not rocket science!" Dan Flynn provided valuable information about the background and bigger picture of olive oil issues in the world. Selina Wang brought the whole group through a brilliant synopsis of the chemical hallmarks of good and bad oil. While in California, we visited Pamela Marvel and Stuart Littell in Capay Valley, whose certified organic oils have been recognized widely and garnered 23 gold medals, a Best in Division, a Best in Class in the half dozen years they have been harvesting from trees they planted. We headed up north and visited with Brendon Flynn and Pablo Voitzuk from Pacific Sun in Tehama County whose "Proprietor's Select" oil has won Best of Show and three gold medals and Tuscan Blend won Best of Class in the high stakes of the New York International Olive Oil Competition. Thanks to Liz Tagami and staff at Lucero Olive Grove, Mill, and shop in Corning where we toured olive groves of every shape and size in the 40 °C+ heat and then did a blissful tasting of their delicious award winning mono-varietal oils inside in the air conditioning, and to Emilio, master miller and former Greek Orthodox brother at Chacewater Winery. Vicki at The Olive Press in Sonoma, who leads dozens of customers every day through olive oil tastings at the grove store, who taught us about her "three pillars of tasting," and we met Pepe the master miller there who oversees day to day operations at the press. Our final CA visit was with Olga from Olica olive oils, who with her partner Bob is a newcomer to the olive business but full of smarts with ambitious plans that have already paid off in producing some extraordinary oils.

The full stories of all of those visits can be found on our blog: World Olive Press, www.worldolivepress.blogspot.com.

Finally, we are grateful to Marcia Grant, the fourth pillar of our Olive Table, who kept us going during early phases of the project. Her position as Rector of Ashezi in Ghana for the last few years led her to turn with bottomless energy and wisdom elsewhere in the later phases of our project.

Financial support through our home institutions of Amherst College and Keene State College for our sabbatical was greatly appreciated. In addition, the Whiting Foundations support was critical to both POH (2013) and RAB (2014) and helped to make this dream a reality.

ZDN is grateful to her family and husband for all their support.

To our beloved daughters on both sides of the globe,
Sarah, Becca and Esin

Contents

The Chemical Story of Olive Oil: From Grove to Table
By Richard Blatchly, Zeynep Delen Nircan and Patricia O'Hara
© Richard Blatchly, Zeynep Delen Nircan and Patricia O'Hara, 2017
Published by the Royal Society of Chemistry, www.rsc.org

Contents

CHAPTER 1

Olive Origins

In this chapter, we start by outlining the history of olive trees and their important role in the development of human civilization in the Mediterranean. Ancient Greek, Roman, Jewish, Christian, and Islamic writings pay homage to the olive tree and describe how critical it is in providing food, heat, and light necessary for survival. The near immortality of the trees connects them with the ancient gods whose own immortal existence is mirrored in ancient trees such as the one shown in Figure 1.1. We will take you with us to excavations from Neolithic villages, where we can see that our ancestors used olive wood in their campfires, and to archeological sites of the bustling Bronze Age city of Klazomenai, where the pressing of olives at a communal olive press used a clever system of rock hewn holding tanks for separating the oil from the fruit liquor.

While we typically think of warm Mediterranean countries whenever olive oil is mentioned, we now understand that olives can be grown throughout the world in the "olive belt" of about 30° to 45° latitude in both the Northern and Southern hemispheres as long as the land is sufficiently dry.[1] In places like Spain, Italy, Greece, Turkey, and Tunisia, the "Old Olive World," new commercial groves with millions of trees and modern presses stand next to family groves of several hundred trees that

The Chemical Story of Olive Oil: From Grove to Table
By Richard Blatchly, Zeynep Delen Nircan and Patricia O'Hara
© Richard Blatchly, Zeynep Delen Nircan and Patricia O'Hara, 2017
Published by the Royal Society of Chemistry, www.rsc.org

Figure 1.1   An ancient olive grove near Yeni Foça, Turkey.

can be many hundreds of years old and village presses that were built at the turn of the last century. In the Mediterranean, many people either own their own groves or know someone who does. Most people who live in the country will have an olive tree or two in their garden. They may have consumed olive oil all their lives without ever having bought any! In general, consumers have very strong ideas about the way things should be done and how olive oil should taste. Modern agricultural practices can sometimes clash with the cultural olive heritage of the Old Olive World. By contrast, new groves and presses in areas like California, Australia, New Zealand, South Africa, and Argentina exist in the absence of that tradition. While some of the earliest European settlers in these former colonies brought olive trees with them, no substantial olive industry developed here until the end of the 20th Century. Today, in these "New Olive World" countries, groves of a million or more trees are harvested and transported within hours to modern facilities where they are processed using state of the art centrifugation methods. Perhaps the vision statement for the South African Olive Association "Old World Ideals with a New World Vision" best captures the contrast.

Today, it is very difficult to create quality olive oil in bulk without a team of dedicated people who plant, prune, and fertilize the trees, prepare the soil, harvest at the right time with the right equipment, deliver the fruit to the press in as timely a way as possible, process the fruit with respect for the natural goodness, bottle the oil with great care to preserve the quality and minimize degradation, perform chemical tests to ensure that the oil meets the highest standards, ship the oil according to accepted protocols, and store and sell the oil with the interests of their consumers in mind. As this book takes you through each of these stages in olive oil production, it will introduce you to a contemporary olive oil expert from around the world whose knowledge, vision, resourcefulness, and enthusiasm is emblematic of the hundreds of thousands of professionals who cooperate to bring the world's best olive oils to your tables. These individuals have helped us to understand the links between history, quality, production, and processing.

Since the soul of the oil is made from the molecules that compose it, each chapter will also feature a particular molecule or molecules that best represent the ongoing development of the olive from grove to table. We will start in this chapter with triolein, the molecule that makes up the bulk of the oil itself and then, in later chapters, introduce you to the important and unique molecules that are present in smaller quantities but are responsible for the wonderful flavors, fragrances, and health effects. Introducing these natural compounds allows us to provide a richer explanation of the techniques of growing and processing, as well as of the impact on humans such as taste and smell, and can be proven to be responsible for the many health benefits of olive oil.

Most of all, we hope the readers will finish with an appreciation for the extraordinary effort required to make a high-quality EVOO (extra virgin olive oil), an understanding of what makes a good quality oil, and an increased resolve to include more of it in their daily life.

1.1 A CULTURAL LEGACY OF OLIVE TREES AND OLIVE OIL

Ancient Islamic, Judaic, and Christian texts refer to the olive tree and olive oil as sacred. Olive imagery has been captured in many cultural traditions[2] and great writers and thinkers from Homer, Hippocrates, Columella, and Pliny all wrote of its almost magical healing and anointing properties. Mustafa

Kemal Atatürk, founder and first president of the Turkish Republic, was an advocate for modern olive production.[3] Even Thomas Jefferson, one of the founding fathers of the USA, believed in the beneficence of the olive tree. Let's see what has been said about the origins of the olive tree and the utilization of its wonderful oil in and out of the kitchens of our forefathers.

1.1.1 An Olive Tree is "...a Covenant Between God and His People"

The first origin story for the olive tree comes to us from the early Judeo–Christian tradition as documented in the 1st Century manuscript on the Penitence of Adam from the *Vita Adam*. At the end of his life, Adam sent his third son Seth back to the Garden of Paradise to request the oil of divine mercy promised to him by God for his own redemption and that of humanity. Instead, the Angel guarding the gate gave Seth three seeds from the Tree of Life (or the Tree of Knowledge). Instructions were given to put the seeds in Adam's mouth upon his death and bury them along with his body. When the time came and Adam died, Seth followed the Angel's instructions and buried his father along with the seeds. A short time later, three saplings (or one tree with three branches) grew from the burial site. Though texts differ on the eventual destiny for the tree(s) at least one story has the trees growing into a cypress, a cedar, and an olive tree – three classic trees of the Middle East.[4] According to one 15th Century text, the wood from these sacred trees grown out of the flesh and bones of Adam later become the rod of Moses and the cross on which Jesus was crucified.[5] Figure 1.2 captures images of this story of the origins of these three trees from a 15th Century Italian fresco, a 15th Century Dutch woodcut and a modern photographic installation in Israel.

So, the olive tree appears as a gift from God to these humans, and represents a promise of his mercy and steadfast love. The Old Testament of the Christian bible makes more than 100 references to olive trees, olive oil, and olive branches. Perhaps most familiar is the olive branch brought back by a dove to the Ark and given to Noah to let him know that the floodwaters

Figure 1.2 Adam's arboreal legacy was three trees, a cedar, a cypress, and an olive tree, grown from the three seeds given to his son Seth by the Archangel Michael to place under the tongue of his father before his burial. Top panel: *Death of Adam*, a 15th Century fresco by Piero Della Francesca; bottom left: Woodcut from *The Legendary History of the Cross* by Veldner 1483; bottom right: photo by Noga Kadman of Olive tree sculpture of Ran Morin, near Kibbutz Ramat Rachel. (Image credits: top image © http://WikimediaCommons/CC-BY-SA-3.0/GFDL; bottom left: public domain Gutenberg Project ebook 46800; bottom right: image © http://WikimediaCommons/CC-BY-SA-3.0/GFDL.)

had receded (Genesis 8:11). With the branch came a promise –
a covenant – that God would never again repeat this total
destruction.

The use of olive oil for religious anointing is a frequent reference as is its use as a source of light against the darkness.

> *"And thou shalt command the children of Israel, that they bring
> thee pure oil of the beaten olive for the light, to cause the lamp to
> burn always." Exodus 27:20*

The New Testament holds a similar symbolic importance for the
olive tree. On the night before he died, Christ retreated with his
disciples to the Mount of Olives and was later captured close to
the local olive press at Gethsemane (Mark 14:26–32). In Romans
11:17–24, the apostle Paul discusses grafting of wild saplings onto
strong olive root stock and uses this as a metaphor for God's cultivation of his people, seeking always the fruit of kindness.

1.1.2   An Olive Tree is "...a Gift of the Goddess Athena"

Greek mythology contains another version of the origin of the
olive tree. Here, it is linked to the warrior goddess Athena, daughter of Zeus.[6,7] The legend speaks of a contest designed by Zeus to
be held for the gods. Each contestant was to prepare a special gift
for the people of a city in Greece and the inhabitants of that city
would review the offerings and choose the most perfect gift. The
god who designed this gift would in turn be honored by having
the city bear his or her name – thus guaranteeing the devotion
and loyalty of its inhabitants for all times. On the day of the contest, two gods presented their gifts to the people of the city. First,
Poseidon, god of the sea, struck the ground with his trident and
out of the earth sprang a horse (some versions say a fountain).
Next, Athena used her spear to strike the earth and the first olive
tree instantly sprang forth. Since the olive tree was capable of
producing light, heat, food, and shelter, the inhabitants of the
city chose Athena's gift. The city, Athens, bears her name even
to this day. Figure 1.3 shows a photograph from the Parthenon
in Athens depicting the contest for naming rights to the ancient
city. An olive tree which, like its creator, is (nearly) immortal
stands close to the corner of the Parthenon today.

Figure 1.3 Athena's gift of an olive tree is the winner in the great contest for the naming of the city of Athens. Top left: model of the west pediment of the Parthenon in Athens by Tilemahos Efthimiadis. Top right panel: 18th Century painting of the battle between Athena and Poseidon by Noël Hallé. Bottom panel: photo of modern day Parthenon with olive tree just below west pediment. (Image credits: top left © http://WikimediaCommons/CC-BY-SA-2.0/GFDL; top right, © http://WikimediaCommons/CC-BY-SA-3.0/GFDL; bottom image used with permission George Courmouzis.)

1.1.3   Olive Oil is "...Liquid Gold" Homer ~800 BCE

Olive oil was a staple in Greek households and in their imaginations. More than 20 references are made to olive related items in Homer's work.[7] In *The Iliad*, he describes some of its non-culinary uses. It was a critical ingredient in the goddess Hera's toilette as she set about preparing herself for her husband Zeus. Meanwhile Aphrodite anointed the body of the slain Trojan warrior prince Hector with scented olive oil. The Odyssey also contains references to olive oil's place as a treasure to be stored with gold and bronze. Perhaps most poignant is Homer's description of the wedding bed made by Odysseus for his beloved Penelope.[8] One leg of the bed is made from a living olive tree, symbolizing perhaps their steadfast and timeless love. In a clever device to test the returning Odysseus – gone for 20 years – Penelope loudly orders a servant to move the bed to ready it for the man claiming to be her husband. When Odysseus shouts out that the bed cannot be moved because it is made of a live tree, Penelope knows the man standing in front of her is truly her long lost husband and they are joyfully reunited.

1.1.4   Olive Oil is "...the Great Healer" Hippocrates ~400 BCE

Hippocrates believed that olive oil was "the great healer."[7,9] He was aware that a topical application of the oil would help relieve pain from skin abrasions and burns and he recommended it be massaged into wounds to help them heal quickly. Extracts of plants steeped in olive oil were used to provide relief from many gynecological diseases, and its use was recommended in curing infections of the ear, nose, and throat. Olive oil itself was massaged into aching muscles and its ingestion was recommended in small quantities to settle an upset stomach and in larger quantities as an emetic. Plants from celery to fennel to St John's Wort to juniper were steeped in olive oil and used to preserve youthful healthy skin.

1.1.5   An Olive Tree is "...the Chief of All Trees" Columella ~50 AD

Columella was a naturalist writer of Roman times whose complete work was thought to be lost to time until a 12 volume combined encyclopedia *De Re Rustica* was found in the library of a Swiss monastery in the 15th Century. Book 7 contains thorough instructions for olive trees with regard to the ground soil preparation, grove locations, pruning, and fertilizing. A complete description of how and when to press olives is given in Book 12, Chapter 50.

"As soon as the berries shall begin to be of different colors, and some of them are already black, yet more of them white (sic), the olive must be gathered by hand when the weather is fair and sifted and cleansed upon mats or reeds spread under them: after they are cleansed, they must be presently carried to the place where the presses stand, and shut up entirely in new frails, and put under the presses, that they may be squeezed as little a while as can be."

Today, these words would be precisely the same instructions given to an olive farmer who desires to make the highest quality extra virgin olive oil. Pick when the olives are partly turned, be gentle with the fruit, and get the fruit to the press as soon as possible.

1.1.6   Olive Oil is "...Common and Universal" Pliny ~70 AD

Pliny the Elder, the prolific 1st Century Roman writer, spends much ink describing the best methods for the cultivation of olive trees, the pressing of olive oil, and the many medical uses of the olive oil itself. His contrast of olive oil and wine is quite insightful.[10]

"It is not with olive oil as it is with wine, for by age it acquires a bad flavor and at the end of a year, it is already old. This, if rightly understood, is a wise provision on the part of Nature: wine, which is only produced for the drunkard, she has seen no necessity for us to use when new. Indeed by the fine flavor which it acquires with age, she rather invites us to keep it. But on the other hand, she has not willed that we should be thus sparing of oil, and so has rendered its use common and universal by the very necessity there is of using it while fresh." (Book XV, Chapter 3, pp. 16–17.)

Pliny also repeatedly recommends: "Do not shake and beat your trees. Gathering by hand each year ensures a good harvest."

1.1.7   An Olive Oil is from a "...Blessed Tree." Qur'an 600 AD

In the Angel Gabriel's revelations to the Prophet Mohammed, the majesty of Allah is said to be like a light that burns, as in Figure 1.4, from the oil of the blessed olive tree. Unlike that oil, Allah needs no fire to provide illumination.

"Allah is the Light of the heavens and the earth. The example of His light is like a niche within which is a lamp, the lamp is

Figure 1.4   Olive oil lamp at the entrance to a catacomb in Lesvos, Greece.

within glass, the glass as if it were a pearly [white] star lit from [the oil of] a blessed olive tree, neither of the east nor of the west, whose oil would almost glow even if untouched by fire. Light upon light." (Chapter 24, Verse 35.)

This verse has been much analyzed by great Islamic philosophers such as Razi (ibn Zakariya al-Razi) in the 9th Century and Avicenna (ibn Sina) in the 11th Century.[11] Their ideas about divine illumination, "Light upon light," and the majesty of Allah use this text as a touchstone.

Dimitri Gutas, Professor of Arabic and Graeco-Arabic at Yale University, has pointed out to us how important scholars of the Islamic golden age between the 10th and 14th Centuries, such as Avicenna and Ibn Qayyim al-Jawziyya, refer to earlier pharmacological knowledge of the time about the importance of olive oil. After referring to the verse above, they write: "Eat olive oil and anoint yourselves with it, for it is from a blessed tree."[12] The ascription of these recommendations to the Prophet Mohammed makes the advice much more authoritative in the eyes of Muslims.

1.1.8 An Olive Tree is the "Most Precious ... Gift of Heaven" T. Jefferson 1787

Before he was President of the United States, Thomas Jefferson travelled throughout Western Europe, in part to explore agricultural markets and opportunities and to try the mineral waters for the restoration of an injured right wrist.[13,14] It was spring in southern France when he first encountered an olive tree in full blossom. From that point forward, he noted in careful detail the places where olive trees were planted. His travelling journals and later letters from Paris described his enchantment with the olive tree and his hopes to establish a grove of trees in South Carolina and Georgia. He dreamt that olive trees could provide ongoing nourishment for the poor – in particular the African slaves whose diet was so deficient. In his writings from Paris, he urged the planting of a tree for each slave born in the New World so that it could provide both oil and fruit. Unfortunately, his experiment never bore fruit, and while he was certain it was because his farm managers were just too unfamiliar with the crop, it may also have been the hot damp climate of the area. From 1787 until his death in 1826, imported olive oil became a staple in his own diet. His yearly food imports to Monticello included four to five gallons of "Oil of Aix" from Aix en Provence, France.[15] While it would take many years for his dreams of lush olive groves providing quality olive oil at low cost to the public, when it was realized, it would be located in California, 3000 miles away from Jefferson's home. Reflecting back on his life, he ranked his introduction of the olive tree and rice to the United States as equal to his writing of the Declaration of Independence.[16]

1.2 DEVELOPMENT OF OLIVE CULTIVATION AND OLIVE OIL PRODUCTION

When did the first olive tree come into existence? When did domestication begin? When was the earliest olive oil production? Archeologists, archeobotanists, phyllogeographers (scientists who analyze gene diversity), and geneticists have all contributed to our understanding of the origins of modern olives and to the close relationship between human civilizations and olive cultivation.

Figure 1.5   Fossilized olive branch found near Santorini Greece (photo credit Jayson Kowinsky: http://www.fossilguy.com used with permission).

Olea oleaster is a near ancestor to the olive tree with small oil bearing fruit that tends to adopt an evergreen bush like stature. As shown in Figure 1.5, fossilized branches of *Olea oleaster* have been found in Santorini Greece and dated back to the 50th Millennium BCE.

Carbon dating is a technique for determining the age of once living organic material by measuring the amount of radioactive isotope of carbon (^{14}C) left in the sample. For archeological purposes, carbon dated results are often good enough to place the samples within historical context. The 5730 year half-life of ^{14}C effectively limits its usefulness to dating objects that existed in the last 50 000 years. Carbon dating of charred wood fragments found next to an early human settlement indicated that by the 45th Millennium BCE our ancestors burned olive wood in their campfires.[17] By the 10th Millennium BCE, early humans harvested wild olives and pressed them to collect the oil.

Controlling the genes of a plant or animal in order to get superior offspring has been a version of genetic modification since ancient times. This process, called cultivation, is at the foundation of human civilizations. The natural world propagates fruit trees from the fertilized seeds of the fruit that are dispersed randomly through the environment and germinate if they happen to fall upon fertile soil. In these cases, the DNA of the new tree has been created by sexual reproduction and is varied. By contrast, vegetative propagation or cultivation is essentially a cloning process by which a cutting is taken from a tree, and a new tree is grown from this cutting. Here, the DNA is identical to the mother tree. The variation in the genetic profile of fossilized or preserved ancient trees gives us a window into the cultivation of wild trees by ancient humans.

Much of the discussion about where olive cultivation first began focuses on the southern and northwestern ancient Levant territories, as shown in Figure 1.6. By the 8th Millennium BCE, there is evidence from phyllogeography that along the Turkish/Syrian border humans experimented with grafting high fruit bearing cuttings onto more robust but poor fruit bearing root stocks.[18]

Figure 1.6 Map of the Ancient Levant (highlighted in red) where olives were first cultivated. (Image: The Levant 3.png. Licensed under CC BY-SA 3.0 *via* Wikimedia Commons.)

Once early farmers encountered a superior fruit with more flesh and a higher oil content they would have taken cuttings of that tree and propagated orchards from this. It is thought that the spread of cultivated olive trees from the Near East occurred first throughout the Levant, including Cyprus, and then further into the western Mediterranean in two lineages, one *via* Libya towards Italy and the other through Morocco towards Iberia.

When cuttings of these trees reached their new homes, it would be likely that they would be further crossed with local wild varieties, leading to more confusion of cultivar identities. From this point forward, the practice of cultivating olive trees by grafting and pressing olives into oil spread from Asia Minor to the rest of the Mediterranean basin. Phoenician sailors brought first olive oil, and then olive tree cuttings and knowledge of best practices for cutting and pressing, to the westernmost Mediterranean basin, including France, Spain, and North Africa.

By the 7th Millennium BCE, the late Neolithic olive oil "industry" had developed to the point where the pressing was done regionally rather than individually, as evidenced from the thousands of crushed pits and pulp remains found in submerged prehistoric sites off the coast of Israel.[19] The remains of jugs and jars containing olive oil dating back to the 6th and 5th Millennia BCE have been reported.[20] The oil is identified by analysis of the remnants by a process known as gas chromatography, which is explained in greater depth in Chapter 6. Analysis of one microliter of sample is enough to detect the length of carbon chains and their levels of unsaturation. High levels of oleic acid (18C chains, with one unsaturation) convinced the team to conclude the sample must be olive oil. In the 3rd and 4th Millennia BCE Minoan culture on the island of Crete had fully integrated olive oil into their daily lives, olive cultivation into their agricultural practices, and the reverence of olive trees into their spiritual lives, as can be found in paintings and artifacts. The great 6th Century philosopher, mathematician, and scholar, Thales of Miletus showed he could predict olive harvests in what is now modern day Turkey.[21] Predicting a good harvest based on several observations of weather and rainfall, Thales reserved the regions presses in advance at a low cost, and then when the time came and the harvest was indeed plentiful, he sold back time on the presses to the farmers at a high rate. It is reported he did this not to make himself wealthy, but to prove the power of the scientific reasoning.

While excavating the ancient Ionian settlement of Klazomenai in the heart of the modern olive region near Izmir, Turkey, archeologists found three deep pits, cut a meter deep into the bedrock and connected by small channels. An additional channel led toward a couple of shallower square reservoirs carved into the rock. Some additional marks and indentations led the archeologists to believe that they were viewing an early industrial olive press dated to 600 BCE. Later, excavation of the site revealed an adjacent waste collection site for olive pits that clearly demonstrated the use of the facility as a regional olive press. Today the site, close to the modern day city of Urla, has been reconstructed as an educational museum with an active press. Figures 1.7 and 1.8 show how archeologists believe the press was operated 2500 years ago.

1.3 WHAT IS UNIQUE ABOUT THE OLIVE TREE?

Turkish poet Nazim Hikmet wrote "you must take living so seriously that ... even at seventy, you'll plant olive trees."[22] In the Mediterranean, long-lived olive trees are considered to be both a link to the past and one's ancestors, and a link to the future and one's descendants. For scholars, the olive tree's incredibly long life span and the amount of influence it has had on the history, culture, and economy is a source of inspiration and amazement. Some go as far as claiming that the domestication of the olives (wheat and wine) is the reason for the emergence of Mediterranean civilization, which is considered to be the cradle of modern western civilizations.[18] The questions are endless. How does the tree live so long? How does it remain fertile through millennia? Can it make those who consume its oil live long as well? Can the longevity of the olive tree give us any hints about extending our own life spans?

Thousand-year old trees – called millennial trees – exist today in places such as Sicily, Crete, Cyprus, and some remote areas of the Mediterranean. One of them is the monumental tree at Kavusi, Crete, shown in Figure 1.9, which is documented to be more than 3000 years old and is registered as a natural treasure. It sits in a secluded glen with a view of the sea. At nearly seven

Figure 1.7   Reconstructed ancient press at Klazomenai, Turkey. Left: workers pushed the four horizontal handles shown at the top of the image to turn the axle that rotated the heavy stones and crushed the fruit to make a paste. Right top, middle, bottom: the paste was packed into fibrous pillows and the pillows were stacked on a wooden table. The oil was separated from the fruit by using the weight of a giant log to press out the oil. The log could be raised or lowered by pulleys suspended from the ceiling. Oil flowed out of the press into a collecting tank carved into the bedrock floor of the press. (Illustrations and photo used with permission from Ertan İplikçi.)

meters high and with a canopy of nearly 11 meters in diameter, the tree is massive, majestic, and stands in a small grove with sister trees of similar vintage. Grafting scars tell us that around 1000 BCE farmers on the island knew enough to take a hearty root stock and graft to it small branches from trees that produced

Figure 1.8   Reconstructed Ancient Press from Klazomenai, Turkey. Left: photo of the three tanks carved into the bedrock floor and used to collect and separate the oil from the water. Right: the central tank collected the golden olive oil and black vegetable water directly from the press, as can be seen in the middle image. An opening in the connecting wall of the adjacent tank allowed the more dense vegetable water that sinks to the bottom to flow in and be drawn off and discarded. Meanwhile, pure oil could be ladled off the top of the central tank and collected into a separate tank. Eventually this purified olive oil would be transferred into clay and shipped all over the Mediterranean. (Illustrations used with permission from Ertan İplikçi.)

abundant fruit. Today, the trunk, which is nearly five meters in diameter, is gnarled, pocked with bore holes that small boulders can fit into, and twisted by the winds and the weather. One cannot help but feel small and insignificant in its presence. The tree has witnessed so much human history: the lives of men long before the birth of Christ and long before colonization movements sent armies from Greece, or Rome, or the Ottomans, or Crusades, and more. And still it lives.

The grove is a 30-minute drive off a major highway up a twisty, rough-and-tumble country lane. For the hour we were there, we were the only visitors. All we could hear were the distant sounds

Figure 1.9   The 3000 year old Millennial Tree in Kavusi, Crete.

of the wind in the trees and the occasional bells of sheep or goats on the hillside. In late December, the tree was heavy with Koroneiki fruit and stood awaiting harvest. To get oil from this tree felt a bit akin to making cider from apples harvested from the Tree of Knowledge. Yet, in its own humility, the giving tree at Kavusi has provided humans with oil from its fruit, light from its wood, and shade from its leaves for over three millennia. Its branches were even harvested and fashioned into a crown to honor the first female winner of the marathon at the 2004 Olympic games in Athens.

1.4   OLIVE OIL BASIC CHEMISTRY: OR WHY OIL AND WATER DON'T MIX

1.4.1   Molecules are Key Players in Quality

What happens to the olives from this tree once they are harvested? How will the oil be separated? What happens to make one oil better than another? To answer these questions, we begin by presenting a way of thinking about atoms and molecules that will make clear such simple things as why water and oil don't mix. In later chapters, we will use this model to explain even much more complicated questions, such as why the olive fruit is making oil in the first place, what telltale signs allow the farmers to know that the fruit is ready for harvest, and why time is of the essence in getting the fruit to the olive press. It will allow us to go with the olive into the olive press and understand how the press separates the oil from the rest of the fruit and how, why, and when olive oil goes bad. A molecular understanding of the oil provides answers to all of these questions and more. Most importantly, the answers will help us appreciate the differences between one oil and the next and to know that the work of producing a high-quality extra virgin olive oil is no accident.

Molecules are almost unimaginably tiny. It would not be possible to see one using even the most powerful light microscope. By the time that we notice them, they will be present in almost unimaginably large numbers. The next time you measure out a tablespoon of olive oil, think about the fact that your spoon will hold more than a billion trillion molecules.

The atoms. The building block for all matter is the atom, and individual atoms connect with each other to form molecules. The identity and precise arrangement of atoms in a molecule determines the substance's utility and properties. Early models used wooden balls and sticks to show the locations of atoms and the bonds that hold them together. If you've seen models of molecules in books or movies, you might recognize the general form of the ball and stick image in Figure 1.10. It is a model of one of the simplest molecules – water, H_2O – with two atoms of the element hydrogen (H) each shown as a small white ball and one atom of the element oxygen (O) shown as a larger red ball. Each atom provides a nucleus that has nearly all of the mass of the atom and also possesses a strong positive electrical charge due to the protons that reside there. For water, the nucleus of the heavier oxygen atom has eight protons and the lighter hydrogen nucleus has but a single proton.

The bonds. If a nucleus was all there was to an atom, two atoms would never come together to form a molecule, but would instead fly apart with tremendous force due to the repulsion of the positive charges of the two nuclei. Fortunately, the nuclei are held together by even tinier particles called electrons. Electrons have a negative electrical charge and so are naturally attracted to the positively charged nucleus of the atoms. Atoms in most of the molecules we will study here are neutral; that is

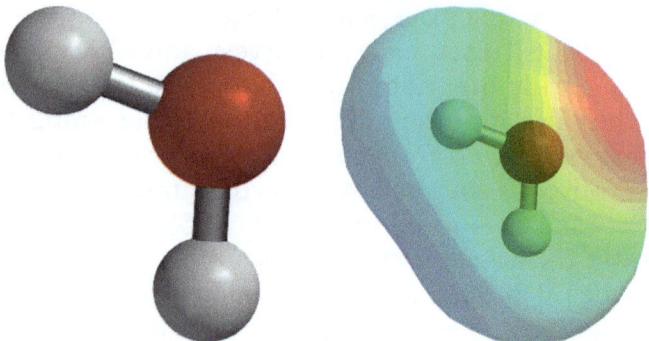

Figure 1.10   On the left, a familiar ball and stick model of water, H_2O, showing a red oxygen atom in the center flanked by two white hydrogen atoms. On the right, a more informative image of a molecule of water showing the ball and stick model in the center surrounded by a surface created from the valence electrons. The surface is what other molecules "feel" when they interact with water. The surface is color coded to show regions that have a high electron density and are slightly negatively charged (red) and others that have a lower electron density and are slightly positively charged regions (blue). Regions that are uncharged are colored green. The prevalence of red and blue on the surface of the molecule indicates that water is polar.

they have an equal number of protons and electrons and so do not have a net charge. Some of the electrons stay close to the nucleus, tucked away from the outside world. The outermost electrons are called valence electrons, and they tend to spread themselves out over the molecule's backbone, usually in pairs. Most of the electrons are found between the nuclei, and the attraction of two neighboring nuclei to the electrons is called a chemical bond. Sometimes, these valence electrons are pulled more toward one atom in a molecule, leaving that part of the molecule with an abundance of electrons and a partial negative charge while another part of the molecule with a shortage of electrons bears a partial positive charge. The water molecule is one such molecule in which the valence electrons are not evenly shared, and so the molecule ends up with an uneven distribution of charge. We say that such a molecule is polar. Because the ball and stick model by itself does not show us the charge distribution or polarity of the molecule, we wish to introduce another way of looking at molecules; that is by considering their "skin" or surface.

Figure 1.10 shows the balls (atoms) and sticks (bonds) standing like a skeleton inside the translucent surface of the molecule. Unless you are a chemist, you may never have seen the skin on the molecule before. This skin represents the outside of the whole molecule, not just the skeleton, and is formed from the outermost electrons. The valence electrons are constantly moving around the molecule at speeds close to the speed of light. The space occupied by a molecule depends mostly on the space explored by these exuberant electrons. While they don't stay in one place for very long, they form a surface that defines the outside of the molecule. It's a dizzying fact that everything we interact with in our daily lives is built this way. For water, you might also notice that the skin is not one color, but is slightly red on one side near the oxygen atom. Remember that the oxygen nucleus has eight protons so, with that much positive charge, it attracts the valence electrons very strongly, thereby creating a partial negative charge, which is shown in the surface model as slightly red near the oxygen. On the other side of the molecule that faces away from the oxygen and lies between the two hydrogen atoms, there is a corresponding blue color, indicating a partial positive charge. This surface model allows us to see that the water molecule is indeed polar and that the charge is not evenly distributed. The significance of this to a single molecule is small, but remember that molecules in our lives come in vast numbers. When two molecules approach each other, the negatively charged part of one is attracted to the positively charged part of the second, making them stick together more tightly. Many of the properties of water, such as its freezing point, boiling point, and density, will be derived from that fact. Whenever possible, we will represent molecules in this text with the skeletal ball and stick model within the skin so that you will be able to visualize the atoms and bonds as well as the true molecular size and shape and the distribution of charge in the molecule.

1.4.2 The Most Important Molecule in Olive Oil: Triolein

Of the many molecules in olive oil, triolein is the biggest contributor by weight and by numbers of molecules. It is part of a molecular family known as the triacylglycerides (abbreviated "TAGs") and each member of the family has a very similar makeup. It is this molecule that is burned by your body to provide the important calories that you need to live – to have your heart beat, your lungs compress and expand, and your brain think. Though it has 224 atoms in one molecule, as food molecules go, it is mid-sized, which means that it is a bit bigger than a simple sugar but much smaller than complex carbohydrates (like starch) or proteins. The TAG molecules break down in your body to form fatty acids that are important sources of energy and building blocks that your body needs.

The atoms in triolein, as shown in Figure 1.11, are of three different elemental types: carbon – element number 6 with six protons shown in black; oxygen – element number 8 with eight protons in red; and hydrogen – element number 1 in white. The ball and stick model on the left shows that this complex molecule of 224 atoms is really made up of four different molecules stitched together: three long carbon chains called fatty acids and one short three carbon molecule, a glycerol, to which the three chains are all connected. The oxygen atoms connect the three chains to the bridge. In triolein, the three fatty acid chains that make up the bulk of the molecule are identical to

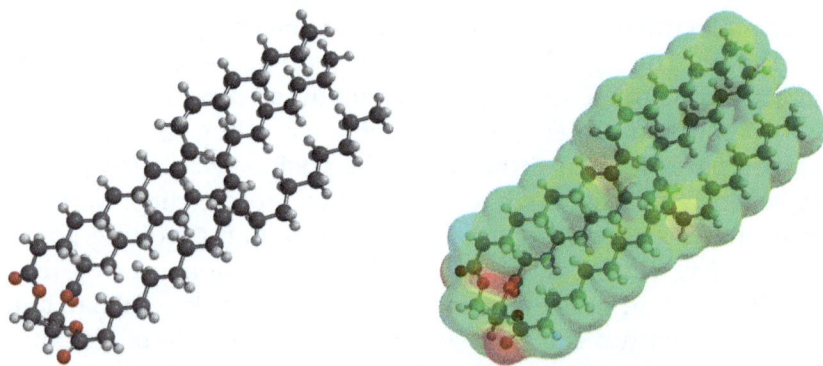

Figure 1.11   On the left is a ball and stick model of triolein, $C_{57}H_{104}O_6$, the most common molecule in olive oil. Oxygen is represented with a red ball, hydrogen as white ball, and carbon atoms are shown as black balls. The more informative surface image on the right shows the valence electron density that surrounds the molecule. Here, a red color shows the surface is slightly negative; blue shows that it is slightly positive and green shows that it is neutral and uncharged. The overall green color of the skin tells us that the electrons are distributed very evenly and the molecule is nonpolar.

each other and are named oleic acid in honor of the olive from which they are mostly derived. Each oleic acid has 18 carbon atoms and a strategically placed concentration of electrons right in the middle between carbons 9 and 10 that make a double bond. Notice the slight kink in the center of the chain that is caused by the double bond. In fact, olive oil has a fairly wide array of fatty acids in addition to oleic acid. Altogether, the fatty acids, including shorter ones (as few as 14 carbons) or longer ones (as many as 20 carbons) are used to use to make the TAGs.

The number of double bonds contained in each fatty acid is a very important property. A fatty acid with no double bonds is called a "saturated" fatty acid; if it has only one double bond, it is called a monounsaturated fatty acid (MUFA); and if it has more than one double bond, it is called a polyunsaturated fatty acid (PUFA). A number of properties and health effects depend on the mixture of saturated, MUFA, and PUFA molecules used to make the TAGs. The net result in olive oil is that triolein is present with a very large number of cousins, which all look very much like it but may have some subtle differences. These differences do not affect the flavor directly, and within the normal ranges of olive oil do not affect its utility. However, there can be a big difference in properties such as shelf life and smoke point. Healthy fats are described as those with less saturated fat, and we are all encouraged to shift our diets to one with less saturated and more MUFA and PUFA. Few oils have as much oleic acid as olive oil, making it an important part of our diet.

If we consider the surface of our triolein, as shown in the model on the right of Figure 1.11, we will notice that, unlike water, triolein is mostly green on the outside. This means that the valence electrons are evenly distributed and we say that the molecule is nonpolar. A small region of very slight positive and very slight negative charge exists on the lower left that is not significant. It may be important to note that the intensity of the color is related to the intensity of the charge imbalance. We said that in water molecules stick together positive end to negative end. Can the molecules of triolein stick to one another if the molecule is mostly neutral? Fortunately, the answer is yes! Weaker forces, called van der Waals forces, exist that make molecules stick together just because they would rather touch each other than be left with no neighbors. While these forces are individually much weaker than the polar forces that cause water molecules to stick together, there are many, many more of them to count when a molecule with 224 atoms approaches another molecule with 224 atoms. So, it can definitely stick together, and that is what we see with the oil.

Now, try to mix oil and water and what happens? As shown in Figure 1.12, the oil, because it is less dense, stays on the top of the bottle and the water sinks to the bottom. This fundamental property is a result of the charged surface of the water molecule finding no complementary charges in the oil molecule, and so it will prefer to stick to itself. In a similar way, the molecules of triolein prefer to be in the neutral environment provided by other

Figure 1.12   Why oil and water don't mix. The structures of oil (triolein) and water (H_2O) on the left are highlighted green (nonpolar), blue (polar with partial positive charge) and red (polar with partial negative charge) to indicate their charge distribution. This difference in the intrinsic polarity of the two molecules results in the spontaneous separation shown on the right of a mixture of oil and water into a yellow oil layer (on the top) and a clear water later (on the bottom) as the polar water and nonpolar oil molecules are driven to avoid each other.

triolein molecules, and so stick to themselves. If you have mixed a salad dressing with oil and vinegar (water) you have noticed that even if you shake it up to mix it, it will naturally separate. This fact – the separation of oil from water – is the foundation for almost all of the separation science that goes into making olive oil and its ability to dissolve other health giving compounds that are themselves not polar.

REFERENCES

1. I. N. Therios, *Olives (Crop Production Science in Horticulture)*, CABI Publisher, Devon, UK, 2008.
2. Z. I. Boynudelik, *Colours of Olive: Olive Image in Art History*, Umur Yayınları, 2011.
3. M. Pabuççuoğlu, *The Agrindustrial Design Symposium and Exhibition: Olive Oil, Wine and Design*, İzmir, April, 2005.
4. E. C. Quinn, *Quest of Seth for the Oil of Life (1962)*, Cambridge University Press, Cambridge, UK, 1965.
5. J. Veldener, W. Caxton, J. Ashton and J. de Voragine, *The legendary history of the cross: a series of 64 woodcuts from a Dutch book*, New York, A. C. Armstrong & Son, 1887.
6. M. P. Morford and R. J. Lenardon, *Classical mythology*, Oxford University Press, 1999.
7. M. L. Clodoveo, S. Camposeo, B. De Gennaro, S. Pascuzzi and L. Roselli, *Food Res. Int.*, 2014, **62**, 1062.
8. Homer, *The Iliad and The Odyssey*, English trans. S. Butler, Barnes & Noble Inc, 2008.
9. Hippocrates, *Hippocrates Collected Work*, Harvard University Press, 1868.
10. Pliny the Elder, *Natural History*, English trans. H. Rackham, William Heinemann Ltd, London, 1962.
11. T. Jaffer, *Razi : Master of Quranic Interpretation and Theological Reasoning*, Oxford University Press, Oxford, 2014.
12. A. Al-Tirmidhi, *Beirut Lebanon: Dar Al-fikr Library*, 1986.
13. T. Jefferson, *Memoirs, Correspondence and Private Papers of Thomas Jefferson, Late President of the United States*, Colburn and Bentley, 1829.
14. *Monticello Keepsake*, http://www.monticello.org/site/research-and-collections/mediterranean-journey-1787, accessed July 2017.

15. *Thomas Jefferson and Olive Oil*, http://www.oliveoilsource. com/article/thomas-jefferson-and-olive-oil-our-first-foodie-president, accessed January 2016.
16. *Twinleaf Journal Online*, https://www.monticello.org/site/ house-and-gardens/thomas-jeffersons-legacy-garden-ing-and-food, accessed January 2016.
17. N. Liphschitz, R. Gophna, M. Hartman and G. Biger, *J. Archaeol. Sci.*, 1991, **18**, 441.
18. G. Besnard, B. Khadari, M. Navascués, M. Fernández-Mazuecos, A. El Bakkali, N. Arrigo, D. Baali-Cherif, V. Brunini-Bronzini de Caraffa, S. Santoni, P. Vargas and V. Savolainen, *Proc. Biol. Sci.*, 2013, **280**, 20122833.
19. E. Galili, D. J. Stanley, J. Sharvit and M. Weinstein-Evron, *J. Archaeol. Sci.*, 1997, **24**, 1141.
20. D. Namdar, A. Amrani, N. Getzov and I. Milevski, *Isr. J. Plant Sci.*, 2014, **64**, 1.
21. G. Sarton, *A history of science*, Xerox University Microfilms, Rochester, 1975.
22. *On Living*, https://www.poets.org/poetsorg/poem/living, accessed June 2016.

CHAPTER 2

The Beginning of a Grove: Planting the Trees

2.1 WHERE ARE OLIVES GROWN?

Most of the world's olive production is found in a Mediterranean style environment: cool, wet winters followed by hot, dry summers. We heard often in Turkey that olive trees grow best when they can see the sea. After seeing many successful olive groves in the middle of a continent, we began to question whether this requirement is for the olive tree itself, or for its owner. Certainly, this traditional association made the study of olive trees much more pleasant for those of us doing the research.

The olive tree is a spreading tree in more ways than one. For centuries, explorers and colonizers from several countries, impressed by the Mediterranean culture and climate, brought and introduced olive trees back in their homelands. By the mid-1500's, the olive tree had been carried to the New World (Mexico and South America). By the 1600's, it had been planted in South Africa, although it was predominantly used for table olives rather than for oil. The 1800's saw a few olive trees brought to Australia and New Zealand. Much more recently, it has made

The Chemical Story of Olive Oil: From Grove to Table
By Richard Blatchly, Zeynep Delen Nircan and Patricia O'Hara
© Richard Blatchly, Zeynep Delen Nircan and Patricia O'Hara, 2017
Published by the Royal Society of Chemistry, www.rsc.org

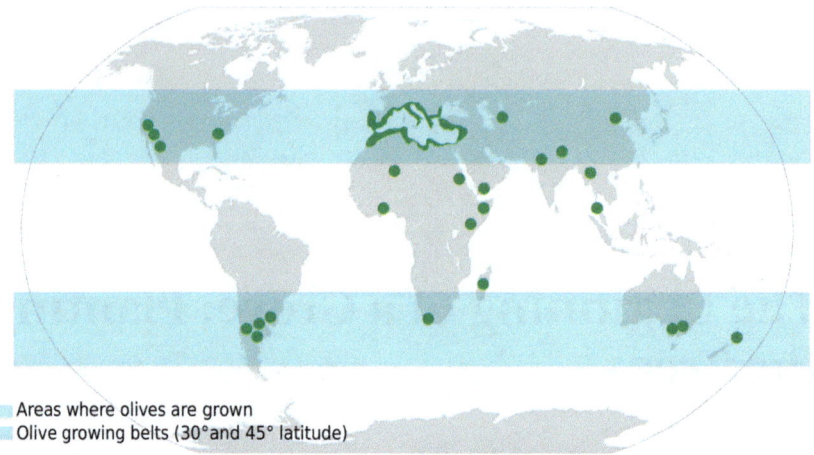

Areas where olives are grown
Olive growing belts (30°and 45° latitude)

Figure 2.1   Geographic distribution of olive production in the world.

it to Japan, China, and Texas, to name a few. Figure 2.1 shows
the two olive belts in the Northern and Southern hemispheres
in which olive production is predicted to be feasible. The green
highlights show that, today, olive cultivation can be found on
nearly every continent and that a few areas fall outside the predicted regions.

2.2   THE OLIVE TREE CAN SURVIVE EXTREME
CONDITIONS...

The most characteristic property of the olive tree is its resilience
to live in relatively good health for a very long time. While considered hardy routinely down to −5 °C, they have been known to survive down to −12 °C for brief periods of time. Researchers were
surprised to see olive trees grow on Mt Judi in Turkey, which has
an altitude of 1157 m, as most olive trees found above 700–800 m
are wild types that yield fruit only sparingly.[1] *Olive Oil Times*
recently ran an article about a German entrepreneur who started
his personal business in Nepal (aka roof of the earth), famous
for its monsoon rains. He planted 2000 trees on eight hectares
between 1400–1900 m altitude and, according to the article,
just profited for the first time after 10 years.[2] The olive trees
that were able to cope with the wet season of Nepal are also able
to cope with no rain under desert conditions. We were told of

Saudi Arabian companies who have decided to grow olive trees in the desert after discovering water at 800 m deep. These companies have ordered seven million saplings to be sent from Agromillora headquarters in Barcelona Spain (personal communication with Xavier Rius).

2.3 ... BUT OLIVES THRIVE IN IDEAL CONDITIONS

Resilience is a virtue, of course. However, greatly increased yield and better flavor may be found if better conditions are provided. To be healthy and productive, it is critical for particular olive tree cultivars to be matched to local climate and soil. The Mediterranean climate, with warm, dry summers and cold winters, is the best natural conditions for the natural growth of olive trees. As we will see in the next chapter, providing water and nutrients can enhance the productivity of the tree and improve the flavor. An ideal climate means that the use of pesticides may be reduced, which is a benefit both to the health of workers and to the economic health of the producers.

In modern times, we are blessed by a very long history of growing olive trees. Thousands of years of careful cultivation have left us with a great body of traditional techniques. Many of these have even been described in print: Columella's horticultural treatise *De Re Rustica*, published in 40 AD, devotes the better part of one book to olive production. Many of the recommendations described there are still valid, as they are based on some very fundamental needs of all trees.

The olive tree, like any growing plant, requires the basic Aristotelian elements: air, earth (soil), water, and fire. Air is, of course, a vital factor in providing for the plant. It contains the carbon source for growth, the humidity the plant works with, the temperature that allows for efficient growth and prevents freezing. Earth supports the plant, and while olive trees are not needy, there are some practical aspects of the site that allow for efficient harvest. Water is vital for any living thing and must be available at the right time. The term fire represents the sun, the source of motive energy. We will discuss the proper balance of these elements below. Of course, for anyone who inherited a grove that is already planted, most of these questions are academic, as the design of the grove is already fixed.

2.3.1   Air

The requirements for successful olive production begin with the correct annual temperature profile, mostly controlled by the latitude. One finds olives grown between 30° and 45° latitude, in both the Southern and Northern hemispheres. There is a minimum temperature below which the olive tree will simply freeze and die. Farmers in Tuscany found this out on a massive scale in a devastating freeze in 1987, during which the bulk of the trees – including some that were hundreds of years old – in the region were killed off by an unusually low and long freeze.

Cameo 2.1   Guido Tinnaci of Casanova Di Tinacci – A Family Farm for Almost a Millennium

2014 was a year of frustrations for farmers in Tuscany – with the production of chestnuts, hay, wine, olives, and wheat all at an all time low. Guido Tinacci, the patriarch of the Tinacci family, met us after his regular job to show us around the farm that had been in his family for almost 1000 years. As the clouds lowered and the drizzle began, he pointed out where he had replanted 500 trees after the frost of 1985. A few of the original survivors stood out in a field now cleared for wheat. The olives remained on the tree, at this point shriveled and pocked. Most of the olives had been damaged by a perfect storm of wet when it should have been dry, dry when it should have been wet, and a resulting infestation of olive fly and a "leprosy" that meant even under perfect harvesting conditions, the oil would be high in acid and have flavor notes that would prevent it from taking the acclaimed extra virgin label. For the first time in almost 20 years, the Tinacci family would have to buy oil from a friend in Puglia, in the south, who were hit less hard. Tinacci was philosophical about his losses. Given that he had a second full time job as an administrator for the municipality, his family would not go hungry. At this moment, somewhat nostalgically, he recalled when olives were harvested after the Feast of the Immaculate Conception, on December 8th, and the oil was sweet and golden. Now, he scoffed, they want the oil to smell like roses and not like olives. Given that he did not own his own mill, Guido was also forced to harvest early because the press would not even be open in December. "It's all fashion" he said, "and we must go along with the fashion." He himself prefers a less bitter oil.

It is a mistake to think that the correct temperature profile is enough, however, as Jefferson found out in his experiment, described in the first chapter. A talented and resourceful farmer, Jefferson was unable to grow olives in the Southern states despite many attempts. He finally realized that the temperature was not the problem in his Georgian grove: it does fall slightly below freezing most winters and is quite hot during the summer. Georgia simply has too much rain and doesn't allow the roots and leaves to breathe, leaving them susceptible to disease. In addition, rain during pollination is a disaster, as it washes away all of the pollen that must drift from one tree to the next on the breeze. Too much rain can also cause problems with any treatment applied to the leaves, which is commonly done with some nutrients and many pesticides.

The best environment has considerable rainfall in the winter, but almost none in the summer. While this is often associated with the Mediterranean region, it is also found in the United States (California), in Argentina, in South Africa (Western Cape), and in Australia, among many others. While olive trees are not particularly more susceptible to wind damage than other trees, it is a consideration, especially at harvest time when the weight of the branches is dramatically higher. In New Zealand we were told it was necessary to plant windbreaks to protect the olive trees from the strong Westerly winds that blow off the ocean and across the island. However, some wind is quite useful for keeping the trees dry and for pollination.

2.3.2   Earth (Soil)

Olive trees are not particularly needy with respect to the soil under them. It can be rocky or sandy, or even clay if the water is managed well. The slope can be flat or precipitous, the covering bare and rocky or a beautiful clover.

Practically speaking, however, the nature of the soil under the trees must be chosen to accommodate the weaker of the species in the equation: the producers who must prune the trees, provide water and nutrients, tend to the pests, and harvest the olives. Depending on the style of harvest that is anticipated, there may be strict limits on the profile of the land. Precipitous slopes allow only quite labor-intensive hand harvesting, and the resulting olives must be carted out carefully. With flatter land, more mechanical harvesting techniques become possible. With the biggest harvesters, producers planning a grove must ensure that the harvesters don't tip over and factor in the space for the machines to turn around, giving a minimum size to the grove and its rows.

One of the most major decisions involves the density of the trees. In ancient groves, one typically sees widely spaced trees. This is the only way that ancient trees can be managed, as the trees themselves have become very large, and these trees were planted using traditional methods (by definition). In most orchards developed before 1990, trees were planted 5–10 m apart in both directions and not irrigated other than by natural rainwater. Some orchards were planted in higher densities, (6 m × 3 m or 7–8 m × 4 m) and were irrigated only if water was available. By the mid 1990's experimentation began with irrigated groves of super high densities (4 m × 1.5 m). Figure 2.2 summarizes the types of plantings and the restrictions common with different densities of trees. Often higher density plantings prove to be more profitable due to use of modified grape harvesters or purpose-built harvesters like the Colossus (discussed further in Chapter 4). Of the roughly 10 million hectares of olive farms around the world, 90% are planted in a traditional method, while approximately 1% are planted using a super high density spacing.[3] A hectare (ha) is a unit of area equal to 10 000 m^2 or about 2.5 acres.

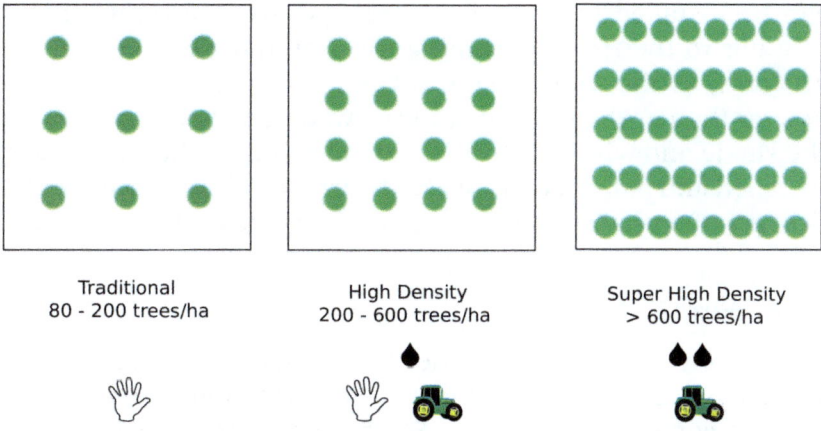

Figure 2.2 Tree densities and traditional practices (irrigation and harvest style indicated).

2.3.3 Water

As we saw above, an olive tree is capable of surviving on relatively little water throughout the year. Requiring about 2 ML ha^{-1} at a minimum (but better at 4–6 ML ha^{-1}), it can live through drought that would kill less hardy trees like almonds. It is a bit hard to visualize this amount of water, but since 2 ML is 2 000 000 liters of water and 1 ha is 10 000 m^2, if this amount were provided solely by rainfall, it would be the equivalent of an annual rainfall of about 200 mm (7.8 inches). If the trees are irrigated, these values allow the producer to find water sources of the appropriate size, as the major irrigation must occur during the dry season. If not, comparison to the annual rainfall in the area, along with the natural tendency of the land to hold water through the summer, can answer the question of suitability of the land. The density of the trees plays an important role in evaluating the water requirements. In a traditional grove, trees must be planted far enough apart that the crowns do not grow into each other, and therefore that the roots do not try to pull water from the same space (the roots spread to a diameter similar to that of the crown).

One advantage of irrigation is that the water can be delivered almost exclusively to the roots using drip irrigation. Once the pattern of planting is set, a network of pipes, pumps and valves can be installed that allows delivery of just the right amount of water to the trees. Designing these systems can be quite an art form, as the water needs at the top of a small hill may be greater than at the bottom, and the water needs will depend on the shade and wind felt by different sections of the grove. Once installed and balanced, however, the irrigation system allows the delivery of both water and nutrients very efficiently.

2.3.4   Fire

Without enough sunlight, none of the other elements are useful. The energy for all of the complex transformations carried out in the tree comes from the sun. Solar energy turns carbon, nitrogen, oxygen, and hydrogen from the air, earth (soil), and water into wood to hold the tree up (and to grow into the soil for water). At the same time, these elements are turned into flowers, capable of turning into olives if pollinated. The driving force comes primarily from the leaves, which are surprisingly complex machines capable of capturing sunlight and converting it to useful chemical potential, and also managing the precious elements of the tree, particularly water in the blazing sun of the summer.

Planning a grove requires consideration of the sunlight available. Groves are typically planted so that the shade of the full-grown trees will only fall minimally on their neighbors. Depending upon the terrain of the grove, the density of planting, and the cultivar, this usually means that the rows have a roughly north–south orientation.

2.4   A CHEMICAL INSIGHT INTO TREE GROWTH – HARNESSING THE ENERGY OF THE SUN

Orientation of a grove to the sunlight and the hours of light that are provided to a tree are key factors that determine the success of a particular grove in a particular place. To grow anything requires energy, which plants get from sunlight. The intensity of the sun's energy amounts to approximately one kilowatt per square meter. Each olive leaf, with an area of approximately 10 square centimeters, then receives about one watt of energy when the sun is shining at full strength. One key step in this process is the absorption of light in a way that allows the use of this energy in chemical reactions. Several pigments are responsible for this absorption.

Surprisingly, there are two major classes of pigments responsible for the green color seen as the olive and the tree grow: green chlorophylls and yellow–orange carotenoids. Like almost all other fruit trees, these pigments may be found on the skin and in the flesh of the fruit in addition to throughout the leaves. These are fundamental to the success of the trees, as they capture the sunlight to drive the conversion of carbon dioxide to sugar.

2.4.1 Chlorophyll

Chlorophyll acts as the antenna for capturing energy from the sun and converting it into chemical energy for use in the plant's survival. This process is called photosynthesis and it is the inspiration for our human civilization's solar energy technology. Chlorophyll must then be found in the youngest leaves in the olive tree, and is found in relatively high levels in leaves throughout the year (hence, evergreen). It is also found in the fruit from its first appearance as a tiny bud, through most of its growth. During the last few weeks before harvest, levels of chlorophyll in the fruit decrease but never disappear.

As seen in Figure 2.3, chlorophyll has a very nonpolar tail, with a moderately polar head atop it. This difference in polarity allows the plant cells to place the molecule precisely into the assembly designed to capture light and turn it into useful chemical potential. In addition, there are some structures visible in this model that are new. In the head area of the molecule, there is a ring-shaped structure represented by red and blue blobs. These are the antenna region of the molecule – special structures optimized to interact with light of the correct wavelength, absorbing the energy. Without these structures, comprised of a ring of unsaturated carbons and nitrogens, there would be no light absorption, and the material would be colorless (like triolein).

Plant leaves in general are optimized for absorbing light to convert to chemical potential energy. The study of photosynthesis is one of the most fascinating branches of chemistry, and while we can provide only the smallest taste of that study here, it continues to provide revelation and inspiration to chemists and society.

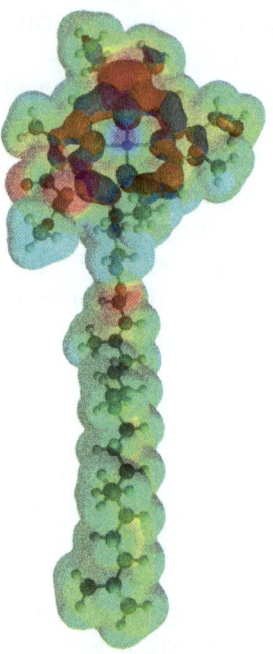

Figure 2.3   A surface model of chlorophyll, the green pigment in leaves. This
model of the molecule shows its polar top where the sunlight is
absorbed and the nonpolar tail that anchors the molecule in the
proper portion of the plant cell.

Let's start by describing the light absorption. The leaf color we
see is a result of the absorption of most of the white light coming
from the sun. This white light is composed of a huge number of
individual particles, called photons, which each have a characteristic energy, or color. When photons of all colors are mixed, we
see this as white light. The ring-shaped part of chlorophyll you
see in Figure 2.3 is optimized for absorbing some of these photons (the blue and red ones). It does this very efficiently, so – as
shown in Figure 2.4 – chlorophyll molecules absorb lots of blue
and red photons (so that their energy can be used to make food
for the plant), and only the green photons are passed along.

Chlorophyll is found in many parts of the olive tree, including the
fruit (especially when green). Chlorophyll is a rather delicate molecule, and does not survive processing, but the immediate breakdown product is very similar to chlorophyll and has a very similar
color. You'll get to meet those molecules in Chapter 5, and find out
how they can help determine the age of olive oil in Chapter 6.

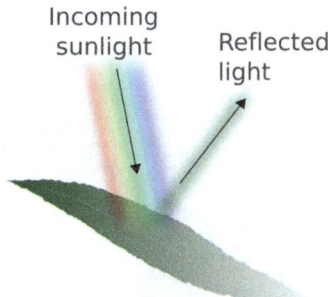

Figure 2.4   Absorption of sunlight by plants. Blue–violet and red–orange-yellow photons in sunlight are absorbed by the chlorophyll pigments in the leaves. The remaining green photons are reflected (as shown) or transmitted. This is why the leaves appear to be green.

2.4.2   Carotenoids

The term carotenoid is a family name for many different compounds, including carotene, lycopene, and lutein. This term bears deliberate similarity to "carrot" as they are the source of the carrot color and coincidentally provide the raw material for vitamin A, used to make the light-capturing pigment in human eyes. As you can see from the structure in Figure 2.5, the molecule carotene also has an antenna-like structure, but one which is long and straight, as opposed to the ring-shaped antenna in chlorophyll. Carotenoids absorb blue light only, making them appear yellow or orange.

2.5   THE PERSONALITY OF THE TREE

We have seen in the introduction that the olive tree has been cultivated by humans for more than 6000 years. The distribution of trees follows much of human history of migration and trade. With millennia of experience selecting trees suited to the local environment, in a wide variety of environments, there are many different varieties of trees to choose from. As in the choice of pet or favorite sports team, this choice can be partly romantic, partly due to tradition, and partly due to a rational approach to suitability to a purpose. In addition, there are some "wild" trees in several of the olive growing regions, including the Mediterranean, of course, and also parts of Africa. Recent studies using genetic analysis of trees from many growing regions have validated both the wide variety and also the migration story of the cultivated plant.[4]

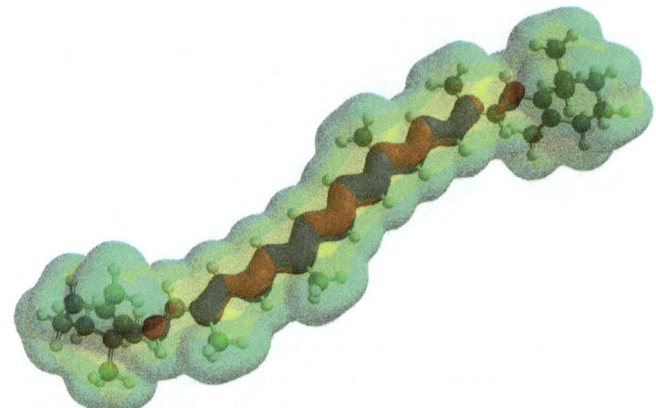

Figure 2.5 Surface model of beta-carotene, the orange–yellow pigment in leaves that extends the light-absorbing ability of the plant.

2.5.1 Cultivars

The word cultivar is thought to be an abbreviation for *culti*vated *var*iety. Cultivars have unique DNA sequences and distinct, uniform, stable characteristics, which are kept through vegetative propagation. This happens only by grafting a rootstock or planting cuttings and is frequently the most productive way to produce new trees.

At least 1250 cultivars are known to be used for olive production around the world, but that number is in flux as there may be many undocumented ones. New cultivars are announced regularly – the most recent being a report of Dr Eleni Melli's discovery of a neglected varietal "Olympia," which grows in the hills near Olympia, Greece.[5] Given that individuals of this cultivar are at least 1500 years old, this is clearly not a new varietal. Olives from these trees produce an oil that is extremely bitter, and with many times the phenolic content of others' oils, it should be an extremely healthy oil. Perhaps ancient Olympic athletes used oil from this cultivar to soothe their aching Olympic muscles.

Nurseries prepare new hybrids regularly to adapt to new demands for super high density plantings and drought resistant crops in new terroirs. Examples of approximately 200 cultivars originating in 26 countries are reported by The International Olive Council[6] and some are shown in Table 2.1.

Table 2.1 Sampling of some common olive cultivars by selected countries (a few representative cultivars and their properties).

Origin	Cultivar	Notes
Spain	Picual	Green with unique pungent and sweet flavor
	Arbequina	Small brown olive, self-fertile
	Abrosana	High yield, compatible with super high density (SHD) planting
	Hojiblanca	More bitter oil
	Manzanilla	"Little apples"; large oval purple green olives, sterile
Italy	Frantoio	Fruity with longer aftertaste, self-fertile
	Leccino	Mild, sweet oils, self-sterile
	Coratina	Mainly near Puglia, self-sterile
	Nocellara del Belice/ Castelveltrano	Large green fruit with mild buttery flavor
	Pendolino	Medium olives, light fruity oil, universal pollinator
Greece	Koroneiki	Small, variegated, high yield, high quality, SHD
	Kalamata	Large with smooth black skin
	Thassos	Small, green, cured on tree
	Olympia	Newly discovered, very bitter oil
France	Picholine	Green elongated medium size, mild, nutty oil
Turkey	Ayvalık/Edremit	Purple fruit, fruity oil
	Memecik	Two silver medals in New York International Olive Oil Competition (NYIOOC) 2016
	Hurma	Naturally cured on the tree
	Delice (not grafted)	Wild bush, tiny fruit, gold medal NYIOOC 2016
Malta	Bidni	Hearty with violet color
Israel	Barnea	High yield, disease resistant, green leaf flavor
Tunisia	Chemlali	Drought resistant, cold hearty, self-fertile, early
USA	Ascolano	Cold hardy variety
	Mission	Planted by Spanish Missionaries, large black

2.5.2 How do the Cultivars Differ?

One of the most important ways in which cultivars differ from one another is in oil content, specifically the profile of fatty acids that is contained in the oil. Another cultivar variation is the flavor of the oil. Flavor is determined by the chemical content of the oil – in particular the antioxidant composition, which adds bitter notes to the oil, and the aroma compounds that smell and taste fruity and give individuality to the oil.

Many other characteristics such as the tree's shape and vigor; its resistance to being blown over; its full grown height; the time it takes to reach maturity; its natural shape; its resistance to cold, pests, disease, and salinity; its fruit shape, size, and maturation timeline; its pollination potential and cross pollination efficiency; and its longevity are all important to the farmer contemplating which cultivars will be best suited to a new plot of land.

Cameo 2.2   Giulio Bertrand, Morgenster Estates, Somerset West, South Africa

Giulio Bertrand had a successful career in textiles and was looking to retire as the century came to a close. A passionate Italian, he fell in love with the beauty of South Africa and decided to buy up the old Morgenster Estate that had been established in 1711. His original vision was to only spend two months a year in the magnificent Cape-Dutch Homestead. Having lovingly restored the homestead, the vision developed and Bertrand believed overseeing the cultivation of the 200 ha of land that were included in the purchase of his "retirement home" would be a pleasant way to spend his retirement. During the restoration period, he became tired of having to constantly fill his suitcases with olive oil from his beloved Italy to last him the year. So he decided to plant the farm to the Bordeaux varieties from which he aspired to make the best Bordeaux blend in South Africa and a few olive trees to make some oil for his salad. It was natural that he decided to try to grow the Italian olive varieties he so loved, on his estate. He imported 2000 bare root plants, selected from all over Italy (from Northern Italy all the way to Sicily) to experiment and to see what grew best on the Morgenster terroir, and today it is thought that the vast majority of trees planted in South Africa since 1994 (some five million trees) can be traced back to the original 2000 trees that were established on Morgenster Estate by Giulio Bertrand.

While trying to identify the major way in which one cultivar differs from another, it is important to say up front that where and how a cultivar is grown will dramatically shape its properties. Whether a tree is robust depends on genetics as well as the climate and soil conditions of the orchard.[3] Robust trees have better resistance to cold, drought, pest, or disease but they tend to be less productive. Some modern cultivars mature quickly and are more delicate, tending to give flowers more easily and bear fruit at an earlier age. Olive oil pressed from a New World *arbequina* olive grown in a high density grove with irrigation will bear only the faintest resemblance to olive oil pressed from the same *arbequina* cultivar grown in Spain in a traditional grove with no irrigation. Both may be delicious and of very high quality, but they will be different. Like wine, vintages are also important: oil pressed in a year with high temperatures and drought conditions will taste more bitter than oil from the same trees pressed in the same press but in a year that has been cool, wet, and rainy.

Perhaps, to the consumer, the most important advice is to taste your oil and to find a taste or set of tastes that you like. Find the oil that best matches what you are looking for in an oil, and realize that you may find yourself buying a different oil to dress your salad than the one you will use to sauté your vegetables or use to accent a soup or other hot dish. Very few people would ask for wine with dinner and leave it at that. Most people will want to specify red or white, cabernet or zinfandel, California or Bordeaux, 2012 or 2003. And we expect the price to vary accordingly. Isn't it interesting that, at least in the US, consumers expect to pay more than $20.00 for a bottle of wine to be drunk at a single dinner in a single night, but complain when that same size bottle of extra virgin olive oil that they will have for months costs the same.

2.5.3 Complete Genome of Ancient Spanish Olive Tree

Recently, geneticists in Spain released a complete sequence of the olive tree *Olea europaea* subsp. *europea cultivar Farga.*[7] The analysis was done on the olive tree shown in Figure 2.6 that is estimated to be 1200 years old and is located near Madrid, Spain. The tree itself has a name, "Santanda, " and like many of its millennial cousins, it is still fruit bearing. DNA was extracted from

Figure 2.6   "Santander" olive tree near Madrid Spain. Photo by Tony Alioto
from Cruz, F.; Julca, I.; Gómez-Garrido, J.; Loska, D.; Marcet-
Houben, M.; Cano, E.; Galán, B.; Frias, L.; Ribeca, P.; Derdak, S.;
Gut, M.; Sánchez-Fernández, M.; García, J. L.; Gut, I. G.; Vargas, P.;
Alioto, T. S.; and Gabaldón, T. Gigascience. 2016, **5**, 29 (used with
permission).

leaves of the tree, but in addition, gene identification was guided
by analysis of other parts of the tree including the roots and
developmental stages such as both young and old leaves, flower
buds, flowers, green fruit, and ripe fruit.

Several notable things have emerged from this analysis. First is
that there are a large number of unique coding sequences – 56 349
genes – almost twice the number of genes compared with humans.
Genes are the pieces of DNA that bear the code for the construction
of proteins that make up the enzymes, signaling molecules, and
structural molecules of the organism. They are located on pieces of
DNA called chromosomes, and these usually come as two copies –
one from each parent. Despite the large number of olive tree genes,
it has the same number of chromosomes – 46 (23 pairs) – as we do.

The telomeric end pieces of the olive chromosomes that protect it from fraying during cell division have been previously reported[8] as being longer than humans – giving us some glimmer of understanding of the extraordinary long lifespan of the olive tree, as we discuss in Chapter 3.

It may be slightly humbling that a tree has more genes than a human, but this is actually common in plant genetics. Plants tend to have many instances of gene duplication – that is, they carry several copies of the same gene. But even when compared with other plants, the olive tree is impressive. The humble potato has 39 000 genes. A plant closely related to the olive tree, *Erythranthe guttata*, or monkey flower, has roughly 32 000 genes that have 75.3% similarity (homology in genetic terms) with the olive tree. What are all those other genes in an olive tree doing? We don't yet know.

Perhaps most striking of all is that this is a genetic map of the oldest living individual ever done. It will give biologists, agriculturists, geneticists, and others years of work to determine how stable the genetic material has been over time, discover how one cultivar varies from another, and thousands of other questions we have yet to imagine.

Ancient robust trees are still being farmed, especially in the Old World, for preservation of cultural heritage. Many families consider their olive trees as part of their family. Artisanal production of such varieties for superior quality and different sensory profiles is becoming popular. It involves hand harvesting and smaller specialized mills that are often costly to run. Many farmers, especially in the New World, are shifting towards high intensity planting of modern cultivars, machine harvesting, and high capacity pressing mills for economical reasons.

2.5.4 Confusion with Names

Estimating the exact number of cultivars is almost impossible because sometimes varieties are named by their local caretakers. For instance, Ayvalık is one of the most common olive cultivars grown in Turkey. The cultivar grown here takes its name from the town of Ayvalık on the West coast of Turkey. Some 300 km north of Ayvalık is another town called Tirilye, which has its own cultivar called Tirilye. In a genetic screening in 2012, these two cultivars were found to be identical to each other.[9]

2.6   HOW DO WE GET NEW TREES?

In Turkish, the wild olive tree (*Olea europea*) is called "delice" or crazy. Crazy olives can grow anywhere from wide, bountiful plateaus to steep, dry hills. They propagate naturally by wind pollination and seeds dispersed by birds. This is called sexual propagation. It is possible to do this with modern olive trees as well, but the barriers are substantial. Without special treatment of the seeds (which we usually call the pits), the success rate of germination is very low. These wild olive varieties are more bush-like than tree-like and the olives are tiny with little oil. Still, they can be harvested, crushed and oil produced from them is quite precious and very bitter.

Most cultivated varieties (cultivars) can self-pollinate and are called self-compatible. Some varieties, however, are unable to pollinate themselves (self-sterile), in which case the farmer needs to plant a different variety with pollinating ability (polli-nator). Pollinator varieties have abundant blooms and produce a large amount of pollen. It is good to have pollinators even in a self-compatible orchard as crossing will induce diversity, which provides resistance in case of abnormal seasonal conditions or discrepancies in the blooming.[10] As most propagation is done by the mist process described below, this focus on pollination is more relevant to growing good olives than producing new trees.

2.6.1   Mist Propagation

Alternatively, trees can be propagated asexually (sometimes called "vegetatively"). If a farmer has a favorite tree producing a superior fruit, he or she can take cuttings of this tree and grow genetically identical offspring to the mother plant. The process requires that the slip from the branch of the desired tree is placed in water or other growth medium and allowed to root. Once the branch has developed roots, it can be planted in soil and grown to full size.

Modern nurseries do this in larger scale, as shown in Figure 2.7. The process of slip selection has been modernized and includes genetic analysis to ensure that the tree is the cultivar you think it is. It is very, very easy to mix up cultivars at this stage of pro-duction. Imagine the farmer who thinks they have ordered sev-eral thousand drought resistant, late bearing trees and instead

Figure 2.7 Thousands of olive tree cuttings being cultivated in the Agromillora Nursery near Barcelona, Spain.

receives and plants young trees that turn out to be a different cultivar that are not drought resistant and produce a sweeter, less robust olive. Since the mistake often cannot be detected until the trees are mature and fruit bearing, this could represent an investment of up to seven years. You can see why the pedigree is important.

The roots develop in a special medium that includes mixtures of hormones to accelerate and stabilize the process. Growing the shoots is a delicate process, as the new roots are not very efficient. Water is delivered through the leaves by frequent misting in a greenhouse (giving rise to the common name "Mist propagation"). Using this technology, companies like Agromillora can take a genetically superior mother plant and propagate millions of micro cuttings.

2.6.2   The Old Fashioned Way: Grafting

Grafting is an ancient art, still practiced today. We have seen a few young olive trees that were grafted, but the bulk of new trees come from mist propagation. However, grafting is a relatively easy process to employ and allows the farmer to easily make the best choice for the two most important characteristics of the tree:

the vigor and water use of the tree, and the fruit bearing qualities. The root stock defines the vigor and water use (including drought and salt tolerance), while the stem stock defines the fruit bearing qualities.

Grafting involves making an open wound in the root stock, often by completely removing the stem above ground. Similarly, a wound is required in the stem stock. If the two wounds are in a compatible shape, they can be joined to make a viable pair, which can grow together. It is important to connect the comparable tissue of each part, so the living part of the tree is continuous from root to top. One of many styles of grafting can be seen in Figure 2.8.

We have wondered frequently about the origin of many activities associated with olive oil production. Who was the first to try grafting, and why did they think this was a good idea? From a naïve outlook, cutting off the top of a tree seems terribly dramatic, and certainly fatal. However, ancient people who watched the trees over many years will have seen separate stems of a single tree grow together into a unified trunk. Watching the branches get close one

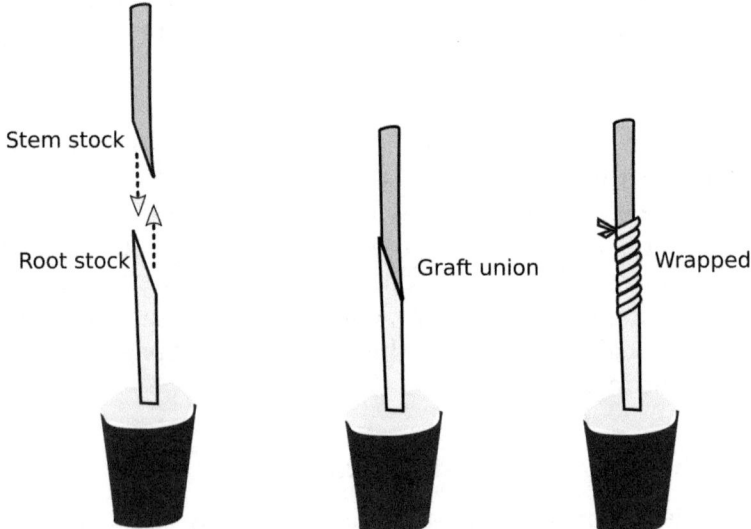

Figure 2.8   Simple grafting of trees. This splice graft is one of several techniques for joining root stock of one tree to stem stock (scion) of a second. This technique is at least 3000 years old. Adapted from Bir, R.; Bilderback, T.; and Ranney, T.; N. C. Coop. Ext. Ser AG-396 Revised 2nd Edition. (Image used with permission T. Bilderback.)

year, rub together to make a wound the next, and make a solid connection the next may have inspired the farmers to try something more dramatic. Of course, there is no historical record of the first grafting experiment, but it does at least mimic a natural process.

2.7   WHAT'S NEXT?

Now that we have trees to plant, earth to plant them in, and air, water, and fire to nourish them, what's next? The next chapter describes the process of growing the trees and producing olives.

REFERENCES

1. *Cudi ve Küpeli'de zeytin ağacı*, http://www.milliyet.com.tr/cudi-ve-kupeli-de-zeytin-agaci/gundem/gundemdetay/30.01.2012/1495747/default.htm, accessed January 2016.
2. *Olive Oil Times*, http://www.oliveoiltimes.com/features/olive-oil-nepal/8791, accessed January 2016.
3. J. L. Harwood and R. Aparicio, *Handbook Of Olive Oil : Analysis and Properties*, Aspen, Gaithersburg, Md, 2000.
4. G. Besnard, B. Khadari, M. Navascués, M. Fernández-Mazuecos, A. El Bakkali, N. Arrigo, D. Baali-Cherif, V. Brunini-Bronzini de Caraffa, S. Santoni, P. Vargas and V. Savolainen, *Proc. R. Soc. B*, 2013, **280**, 20122833.
5. *Olive Oil Times*, http://www.oliveoiltimes.com/olive-oil-business/europe/olive-oil-innovations-discoveries-break-throughs-ancient-olympia/51812, accessed July 2016.
6. *Olive Nursery Production and Plant Production Techniques*, http://www.internationaloliveoil.org/projects/index.htm, accessed January 2016.
7. F. Cruz, I. Julca, J. Gómez-Garrido, D. Loska, M. Marcet-Houben, E. Cano, B. Galán, L. Frias, P. Ribeca, S. Derdak, M. Gut, M. Sánchez-Fernández, J. L. García, I. G. Gut, P. Vargas, T. S. Alioto and T. Gabaldón, *GigaScience*, 2016, **5**, 29.
8. L. J. Fick, G. H. Fick, Z. Li, E. Cao, B. Bao, D. Heffelfinger, H. G. Parker, E. A. Ostrander and K. Riabowol, *Cell Rep.*, 2012, **2**, 1530.
9. A. Ipek, E. Barut, H. Gulen and M. Ipek, *Sci. Agric.*, 2012, **69**, 327.
10. C. Peri, *The Extra-Virgin Olive Oil Handbook*, Wiley Online Library, 2014.

CHAPTER 3

The Tree Through the Year

3.1 WHAT DO OLIVE TREES AND FRUIT LOOK LIKE IN NATURE?

3.1.1 What Does an Olive Tree Look Like?

The typical cultivated olive tree, an evergreen plant, has leaves that are darker green on the top and silvery grey on the bottom, as shown in Figure 3.1. On a breezy day in the Western coast of Turkey, a mesmerizing color show of the shivering leaves alternating between green and silver will impress a visitor. Of course, it bears a fruit that starts off as a tiny green berry, which grows larger during the summer, but then changes color through the fall, usually ending in a dark purple.

The wild olive tree looks quite different from the cultivated one. In fact, the wild olive tree isn't even a tree, but a bush. Its leaves are small and thorny and the wild olive fruit is so tiny that it would yield only about a quarter of the oil obtained from cultivated olive fruit.

Cultivated olive trees do look like trees with a crown and trunk, as shown in Figure 3.2. If left unattended they may grow

The Chemical Story of Olive Oil: From Grove to Table
By Richard Blatchly, Zeynep Delen Nircan and Patricia O'Hara
© Richard Blatchly, Zeynep Delen Nircan and Patricia O'Hara, 2017
Published by the Royal Society of Chemistry, www.rsc.org

Figure 3.1   Olive grove in Turkey.

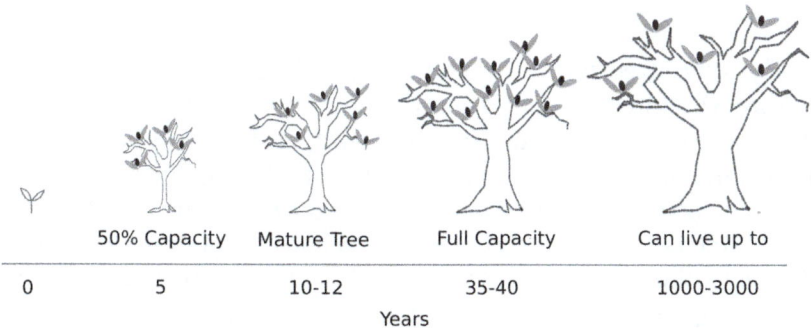

Figure 3.2   Timeline for growth of an olive tree.

up to 10 meters high,[1] though farmers usually keep their trees pruned between one and two meters high for ease of harvesting. According to a report published by the International Olive and Olive Oil Council, an average olive tree will take up to five years to start producing fruit, is considered mature at about 10 years old, and reaches maximum performance by about 35–40 years of age.[2] Trees in newer high density plantations developed by cloning existing cultivars and selective breeding may reach maturity and maximum performance even earlier.[3]

Figure 3.3   Two examples of millennial trees. The first, a 1500 year old olive tree near Yeni Foça, Turkey, (left) and the second, the 3000 year old Monumental Olive Tree (right) of Karvus on the island of Crete that is a national treasure. Left image: Reprinted with permission from Blatchly, R. A., Delen, Z., O'Hara, P. B., *J. Chem Ed.* **91**(10), pp. 1623–1630. Copyright © 2014 American Chemical Society.

The olive tree is one of the longest living productive species. Figure 3.3 shows some of the millennial trees we have visited that are said to be more than 1000 years old, and yet still bear fruit. It is not uncommon to hear from locals in the Mediterranean that trees in their region are two or three thousand years old. The core of the tree is often hollow, either due to natural rot or to the habit of actively hollowing out trees to gather firewood or to provide a space for storage or even as a refuge from the summer heat. Because of the hollow core, and the multiple centers found in many trees, it is often not possible to determine the exact age of olive trees by counting growth rings (which scientists call *dendrochronology*). For an approximate age, it is common to use the diameter at breast height (DBH). A Spanish group tested 12 ancient trees and used the following formula, Age = 2.11 × diameter (cm) + 88.93, to estimate the maximum age to be 627 ± 110 years.[4] These formulas are likely to be specific to a certain group of old trees. Scientists more commonly prefer carbon dating (as explained in Chapter 1) if they find any surviving ancient part of the olive tree.

One reason for longevity is the genetic make up of the tree. The strands of DNA making up the chromosomes in any living species have caps protecting the ends of DNA strands from fraying during cell division, which are called telomeres. It has been suggested that telomere length may be correlated with life span. The longer the telomeres, the more protective they are, and the longer the typical lifespan of the organism.[5] Olive trees have longer telomeres than humans, which may explain their longer lifespan.

Part of the secret to the olive tree's long life may also be a part of the tree known as the ovules, not to be confused with the ovules in the reproductive parts of the flower. These are characteristic swellings of older trees (perhaps a century or more old) at the collar level or where the root joins the trunk.[6] They are commonly referred to as yumru in Turkish, also radish or onion among Turkish locals because they are underground. The ovule, like the hump of a camel, stores water and nutrients, making the olive tree heat, drought, and salt tolerant. Ovules survive below freezing temperatures or above bearable heat (for instance during fires) providing a fresh shooting for the regeneration of the tree. The young shoots continue to grow from the ovule every year but need to be pruned to prevent the tree's reversion to a bush like shape.[7]

Olive tree roots have high plasticity. Their growth pattern depends on the availability of water, slope of the terrain, and the type of the soil but they grow at a depth of 25 cm to 60 cm, which is rather shallow. This is why extensive tilling is not advised. The root system doesn't go deep but instead expands each year, maintaining the same diameter as the crown through the life of the olive tree.

3.1.1.1 Groundcover in an Olive Grove. After tree age, perhaps the single largest variant we saw as we visited groves around the world was the groundcover, or the growth underneath the olive trees in a grove – some examples of which are illustrated in Figure 3.4. Groundcover is determined by the season, what the climate allows, what the needs and resources are, and what the culture dictates. The groundcover is important for preserving moisture in the soil in arid regions, protecting the shallow root system, for discouraging pests in other areas, providing nutrients in a third, or essentially nonexistent in many places. In addition to

Figure 3.4   Examples of groundcover for olive trees around the world. Left: clover in Greece; middle: dried grass in South Africa; and right: green grazing grass in New Zealand.

mulching the trees to preserve water and nutrition, addressing the aesthetics of a grove, or the convenient recycling of manure and straw from a local horse farm, the groundcover strategies often reflect the creativity and connections of the farmer. There simply is not much money available to provide groundcover, given the other expenses in a grove.

3.1.2   What Does the Growing Olive Fruit Look Like?

The fruit growth begins with fertilization. The fertilized ovary grows into seed produced from the outer wall of the ovary and surrounded by an intact fleshy outer covering. The process is fueled by sugars synthesized in the leaves and transported to the fruit. Later in the season, sugars are also synthesized in the green olive itself (shown in Figure 3.5). In the ripening process, plant hormones cause final hardening of the seed (also known as stone or pit), increase in the pH (sour acidic to neutral), breakdown of starch into sugar, softening of the flesh as pectin levels decrease, and degradation of some large organic molecules to smaller aromatic molecules.

3.2   THE OLIVE GROVE THROUGH THE SEASONS

Like most plants that produce an annual crop, there is an annual cycle of activities, needs, challenges, and opportunities to which the producer must be attuned. While in New Zealand, we often

Figure 3.5   Growing olive fruit, Turkey.

heard a line quoted from a guide on olive production stating that oil production is the easiest of tasks: "All you have to do is to sit on your porch, drinking wine, while the oil drips into the bottles." This quotation was always accompanied by either a wry laugh, or a dark look indicating what might befall the author were he to appear on that porch. In truth, production of high quality olive oil is a year-round activity with plenty of challenges. Fortunately, a good harvest is also a great reward, at least in satisfaction.

Olives are harvested in late fall and early winter: roughly September–December in the Northern Hemisphere and March–June in the Southern Hemisphere. In Figure 3.6, we see some of the activities, as well as the color and size of a representative olive through the seasons.[8] Naturally, the months associated with each season differ between the Northern and Southern hemispheres, and there is quite a variation in the length of the growing season. This growing season can be divided into five stages.

The first stage, in the spring, is flowering and development of pollen, followed by fertilization of the female part of the flower. The new seeds (which will become the olive pits but are very soft at this phase) grow quickly for about a month, covered by a thin

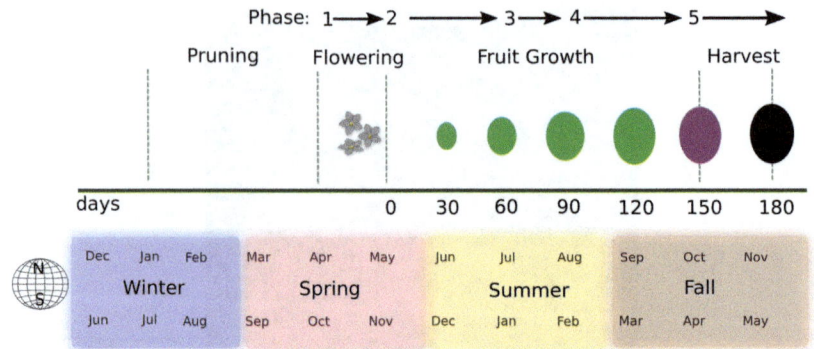

Figure 3.6   Cycle of the olive tree through the seasons.

skin (Phase 2, Figure 3.6). After seed development, the focus of the growing olive turns more to the development of flesh. Seed cell division slows down and a tough shell forms over the pit. Flesh cells divide quickly and oil content increases, adding to the fruit size and weight (Phases 3 and 4, Figure 3.6). Finally, on a signal that comes from day length and temperature, ripening begins. The fruit softens as the color changes from green to purple then to almost black.[8]

The olive farmer must worry about protecting trees against many pests that can reduce a harvest to nothing or, worse, cause a fatal infection in a tree, grove, or region of olive farming. When one considers insects alone, there are more than 100 insects known to feed on the olive tree in the Mediterranean region, but only a handful can be considered harmful pests that are widespread and can cause a significant impact on harvest. Some have simple remedies and the proper application of a pesticide or pest management can address the issue. Others cause apocalyptic scenarios of destruction and cause devastating economic losses in olive growing regions.

We will discuss the activities, challenges (such as nutritional needs, pests, or climate conditions), responses to each challenge, and some of the science behind the decisions that are made. This will not be a comprehensive guide, so as complex as this may look, people who are actually in charge of a grove have a lot more to worry about. We would like to take the opportunity to remind them that massaging the face with olive oil is a good way to remove worry lines (Chapter 9).

3.2.1 Winter

"He who ploughs the olive-grove, asks it for fruit; he who manures it, begs for fruit; he who prunes it, forces it to yield fruit."
Old proverb[9]

Pruning is said to be one of the four most important factors of tending olive trees. The goal of pruning is to remove branches which are dead, less productive, or interfering, so as to maximize the health and production of the trees. This sounds easy, but as you'll see, production is not just getting the tree to grow fruit, but also allowing the farmers to harvest the fruit. Some tree shapes are designed to fit mechanical harvesters or to allow workers easier access to the highest branches. A tall mechanical harvester can manage a tall tree, but that tree can't be too wide. Hand harvest is best done from the ground, which means that the lower the tree, the easier it is to harvest. It is also important, as it is with all fruit trees, to provide open space for airflow and sunlight, which provides more energy for production.

In addition to pruning the fruit-producing top of the tree, shoots or suckers that grow from the base of the tree must be removed to maintain the single trunk shape of the tree. Also, because these grow from the rootstock or ovules, they can be a different variety if the tree was grafted, and compete for nutrients and water with the fruit producing parts of the tree.

Pruning and use of fertilizers to supply the soil with nutrients can change the yield of the crops. The basic principle is to allow the green healthy leaves to be exposed to sunlight so that they can efficiently photosynthesize and produce the necessary nutrients for the development of the tree and growth of fruits. Pruning is a very labor-intensive activity and requires quite a bit of experience and training to do well. One of the more efficient operations we saw, with over 1 million trees, used teams of pruners who averaged about one minute per tree (go ahead – you try it!). Even at that remarkable pace, it would take about 100 workers about four work-weeks to finish the pruning.

In some countries, domestic or wild animals live in the groves with the trees. Some species will eat the tender young shoots – the animals of choice in New Zealand are the sheep seen in Figure 3.7; grazing happily provides nutrients to the tree through their manure.

Figure 3.7   Sheep and trees near Hawke's Bay, New Zealand.

3.2.1.1 Diseases.   Winter can be a time of pre-emptive strikes against disease. Pruning is the first step in fighting disease, as an open tree with no dead or diseased wood is healthiest. Application of dilute solutions of copper ions to the leaves reduces the chances of some fungal infections later in the year, and the time just after pruning is a good opportunity for that activity. These copper formulations often contain copper sulfate, which when solid, forms beautiful blue crystals that give rise to the Turkish layman's term *the eyestone* (also blue and similar to the famous "evil eye"). The copper is usually mixed with lime to make a solution known as Bordeaux mixture. The name comes from the Bordeaux region of France where use of copper as a fungicide was first discovered in the late 19th Century in fighting fungal infections in vineyards.

The entire surface of the plant needs to be covered by the mixture evenly. As the solution dries, copper ions persist on the plant surface. Slow and gradual release of copper ions as the plant gets wet ensures reduced risk of phytotoxicity.[10]

What diseases do the growers worry about? They fall into two broad categories: fungal and bacterial. Infections in trees are like those in humans – they can occur by various mechanisms (insect bites, contact with other trees, or water or wind-born transport of infectious material). Healthy trees are less susceptible, but some cultivars are just more sensitive to disease than others.

Local conditions can make the disease worse – for example, a wet environment makes trees more likely to get a fungal disease if infected. While some diseases seem new, the problem of olive diseases is very old. Theophrastus, the 4th Century BCE Greek philosopher, described olive knot disease at least two millennia before it was classified as a bacterial infection.[11]

Fungal infections include anthracnose, peacock spot, root rot, and Verticillium Wilt of the Olive (VWO). Anthracnose is a fungal infection found throughout the world and invades all parts of the plant – from buds to leaves to roots. It leads to low fruit yield and poor quality fruit.

Peacock spot or peacock eye, as shown in Figure 3.8, is caused by a fungus called *Spilacaea oleagine,* commonly seen around the Mediterranean basin but also in the New World growers such as California, Australia, South Africa, and South America. The most distinct symptom of this disease is ring-shaped spots (Peacock eyes) and falling of leaves starting at the end of fall, continuing until the spring.

VWO has become one of the most threatening diseases in Mediterranean orchards in the last 25 years since its first report in 1946 in Spain. It is a disease of the water conducting (vascular) tissues caused by a soil-borne fungus called *Verticillium dahliae.*[12] Leaves might wilt and lose their green color due to loss of

Figure 3.8   Olives with very mild peacock spot found in South Africa.

chlorophyll; a process called chlorosis. Treatment for VWO may require destruction of the grove, followed by intense soil heating.[13] Another interesting combat strategy is use of biological control agents, such as sawdust and broccoli debris (personal communication with Dr Cahit Tunç). These have been shown to change the soil environment in favor of antagonistic microflora.[12] Broccoli contains glucosinolates, a group of sulfur-containing compounds with a pungent taste similar to mustard. At high concentrations they are toxic and act as natural fungicides.[14]

Cameo 3.1   Cahit Tunç, Born Under an Olive Tree

Cahit Tunç, a Turkish man from the small town of Geyikli near the ancient city of Troy, is also an international scholar of olive production, with an advanced degree in agricultural engineering. As he accompanied us to the small coastal town of Küçükkuyu, he told us that, after being born under an olive tree, he decided it was his destiny to focus his PhD project on the olive tree. We all learned so much from this humble man who also happened to be the Chief of Research and Development of the largest olive oil cooperative in Turkey. He showed us his rather unorthodox method of drastically pruning older trees and how to harvest so as to NOT destroy fruit buds for next year's harvest. His "green" method of preventing the spread of contagious fungus *verticillium dahliae*, was to spread wood dust and plant broccoli (not cauliflower!) around the roots. Dr Tunç just retired from his job as the Chief of Research and Development but continues to guide farmers in the region about organic farming through his organic farming consultancy company.

3.2.1.2 Bacterial Infections. Bacterial infections include the olive knot disease, Mediterranean black scale, and *Xylella fastidiosa*, usually referred to as just Xylella. Olive knot, as shown in Figure 3.9, is called "olive branch cancer" in Turkish, after its tumorous symptoms on branches. The bacteria are dispersed *via* wind, rain, insects, or human activities (such as too-vigorous harvesting with sticks or unsanitary pruning practices) and enters the plant through wounds. The bacteria thrive under warm and wet conditions. Mediterranean black scale (*Saissetia oleae*) attacks leaves and young branches. Contrary to its suggestive name, it most likely originated in South Africa and currently is spread throughout the world. Severe infections can lead to a several year absence of fruit production. Antibiotic treatment can be helpful in killing the black scale – to prevent spreading, care should be taken when young trees are transferred.[15]

Xylella is one of the world's deadliest plant bacteria and has been making headlines since early 2015 because of the outbreak of a new strain among olive trees in the Puglia region of Italy. Approximately 850 km^2 have been put under quarantine and the EU has recommended the destruction of more than one million olive trees or about 10% of the olive trees in the region. Infection of a tree by Xylella leaves them barren – denuded of leaves. At this writing, there is no cure for the disease. Since the bacterium

Figure 3.9   Olive knot observed on an olive tree in Tuscany, Italy.

is transported by insects, the likelihood of the infection of new areas is quite certain unless drastic measures are taken. Farmers in the Puglia region are understandably quite upset about the sacrifice of their groves and many have refused to cooperate with the EU recommendations. Protests and a desperate search for other options continue.[16,17] Scientists studying the disease in search of a cure have been placed under extraordinary pressure, and even accused of being the cause of the infections.[18]

Plant disease, a fact of life in any agricultural venture, is summarized for the olive tree in Table 3.1. The issue of disease in olive trees can be much more emotional than in many other crops, as the trees have such a long life span and are often a revered part

Table 3.1   Summary of various diseases, treatments, and outcomes for the olive tree.

		Diseases			
Season	Type	Name	Symptom	Prevention/ cure	If not cured
Winter	Fungal	Anthracnose	Rotting fruit	Bordeaux mixture	Low fruit quality
		Peacock spot	Ring shaped spots, leaves fall at the end of fall	Bordeaux mixture	Low fruit quality
		Verticillium wilt (VWO)	Leaves wilt, chlorosis	Heating soil or change soil microflora.	Destruction of grove
	Bacterial	Olive knot	Tumors on branches	Prevent infection by sterile equipment	Low fruit quality
		Mediterranean black scale	Marks on leaves, young branches	Antibiotic	Low yield or absence of fruit
		Xylella	Leaf scorch	No cure	Quarantine, destruction of groves
Spring	Pests	Olive seed wasp	Lays eggs in seeds		Low fruit quality
Summer		Olive fly	Lays eggs in olive	Traps	Low fruit quality

of the family. Harsh measures, such as the state-ordered destruction of groves in Italy, are fiercely contested.

3.2.1.3 Nutrition. Nutritional and water needs of the trees are at their lowest in the winter, not least due to the low temperatures of the season. In many of the Mediterranean climates, it is also quite rainy during the winter, making this a time to fill reservoirs, rather than empty them for irrigation if that's practiced. Growth of branches and leaves is slow, and there are few active pests. This is a time of reflection on the past harvest, and preparation for the upcoming one.

3.2.2 Spring

Spring is the awakening of the growing plant, as the days get longer and warmer. The intake of basic nutrients starts in earnest and the new stems and leaves begin to grow. Spring starts the need for general nutrients, as the development of new leaves and fruit begin to accelerate in the warmer temperatures and longer days. Any tree needs the basic elements just as we do: carbon, oxygen, nitrogen, phosphorus, and potassium in particular. Of course trees, unlike people, can take their carbon supply from the air in the form of carbon dioxide. Nitrogen, phosphorus, and potassium do not come naturally from the soil but often must be supplemented. This can be done by sprays in a process called "foliar application" (meaning spraying it on the leaves), but is most convenient through a drip irrigation system.

3.2.2.1 The Need for Fertilizers. The resilience of olive trees is a source of pride for farmers in the Old World although productivity of the trees may be reduced as a result of insufficient nutrients. Natural or synthetic, without sufficient fertilization, soil quality deteriorates, reducing the nutrient levels of the soils and the fertility of the orchards. For those who want to avoid synthetic fertilizers, a possible practice is to leave the pruned wood and plant left overs on the soil as a natural source of nutrients. Another possibility is to let animals roam through the grove and leave manure. Such practices are seen the Mediterranean with sheep and goats or in Australia and New Zealand with kangaroos and sheep. These animals can also eat the young shoots, helping farmers to keep the trees in shape. While natural sources of

Table 3.2   Recommended nutrients for olive trees.

Essential nutrient	Function	Signs of deficiency	When to apply	Amount per tree
Nitrogen (N)	Synthesis of proteins, genetic material, new tissue	Pale color, slow growth	Just before the greatest uptake – early spring to early summer	1 kg
Phosphorus (P)	Important for energy molecules ATP, ADP	Slow growth, chlorosis. Best diagnosis by soil analysis	Only if needed	0.5 kg
Potassium (K)	Stomata control, internal water balance	Dead tissues at the tip of leaves, curved downward. Best diagnosis *via* leaf analysis	Regularly by fertigation system in the spring and entire growth season	1 kg
Boron (B)	Cell wall development, flowering	Chlorosis, low quality of fruit	Foliar spray immediately prior to flowering	0.025 kg

nutrients have many benefits, it is often necessary to use synthetic ones as well. The balance will often depend on local practices and the size of the grove. Table 3.2 gives a summary of some common nutrients, discussed below, and the consequences of deficiency.[19,20]

In many areas, it is common practice to decide how much of these nutrients to apply to the orchard intuitively, according to visual cues. However, this practice is risky as the over-accumulation of the nutrients can be detrimental to the quality or quantity of the produce. Olive orchards can have a patchy distribution of nutrients after applying a homogeneous amount of nutrients throughout the orchard,[20] leading to possible overdose. Leaf analysis at the right time could provide valuable information for both increasing the productivity and quality as well as reducing waste and damage to the local environment. In large groves, this managed delivery is combined with irrigation, by mixing soluble fertilizers in the irrigation water, in a process we heard referred to as "fertigation."

3.2.2.1.1 Nitrogen (N for DNA). Nitrogen is an element required for the production of proteins and the genetic material DNA, among many other plant components. As such, it is required in substantial levels for all growing plants that need to make new tissue. In addition, sufficient levels are crucial for the flowering of the olive plant. At ideal nitrogen levels, plant ovules live the longest so there is more time for fertilization. At low levels of nitrogen, flowering and shoot growth decreases. On the other hand, if the levels are too high, nitrogen compounds accumulate in the fruit, negatively affecting the fruit quality.[21] In low density planting, with a well-developed groundcover with natural animal fertilizer, the soil readily provides plants with the necessary nitrogen. Higher density and more intense production will deplete the soil more quickly, requiring some supplements. Because plants can't use nitrogen from the air, there must be another source. An intermediate solution, which we observed in Crete, is the use of clover, which is a nitrogen-fixing groundcover. In higher density production, farmers can provide these nutrients in the form of fertilizers, most commonly ammonium ion (NH_4^+) or urea (NH_3COCH_3), or sometimes nitrate ion (NO_3^-).

3.2.2.1.2 Phosphorus (P as in ATP). Like nitrogen, phosphorus is a fundamental nutrient and is present in many molecules needed for growth. It is a key part of DNA and is important for many molecules involved in the production of energy, such as ATP. An adequate level of phosphorus is important for a strong root system and also ensures the correct timing of flower production, fruit set, and total fruit production. Just like nitrogen, low levels of phosphorus decrease the flowering.[22] Phosphorus is taken up from the soil by the plant roots, in the form of phosphate ions (mostly dihydrogen phosphate $H_2PO_4^-$).

3.2.2.1.3 Potassium (K, the Gatekeeper). While important, potassium deficiencies do not have the same effects as nitrogen or phosphorus. Only prolonged and severe potassium deficiency reduces flower intensity and productivity.[22] Potassium is generally added through irrigation water if possible, but can also be applied through the leaves. It regulates internal water balance. This has to be timed correctly, with application in the spring much more effective than in late summer in non-irrigated trees. The reason for this gives us a window into why the olive tree is such a wonderful dry-climate plant. After a period of low water availability, the leaves automatically close the stomata, or openings to the atmosphere, tightly. This puts the tree into a water conservation mode, but also blocks nutrients from entering through the leaves.[21]

3.2.2.2 Flowering. The highlight of the spring for any productive species is the development of the fertilized seed. This is an extremely complex process requiring the rapid growth of flowers (an exquisite structure on the tree), the development of and release of pollen, and the capture of pollen by neighboring trees. The olive tree requires help only from the wind for pollination. However, many cultivars are not self-pollinating, meaning that there must be a mixture of cultivars in most groves.

3.2.2.2.1 Boron (B for Flowering). Boron is probably not the first element that comes to mind when talking about plant nutrients, especially for those who aren't farmers. Yet, even some farmers pay little attention to this vital nutrient if they are not watching for the signs of deficiency or performing analysis. The olive needs the most boron of any tree, more even than apple, which is the next most needy.

Boron is vital to the production of strong cell tissue. It combines with strings of sugars called pectins to crosslink them around cellulose fibers to make the strong latticework necessary to provide the strength needed to support the weight of a tree. Similar strength is needed to preserve the shape and function of delicate functional tissue like flowers, which produce the fruit if successfully pollinated.

Needless to say, boron in the tree has a crucial role in the production of flowers and fruit.[23] Quickly developing a complex structure like a flower without enough boron is impossible, and the flowers that develop are not functional. Boron-deficient trees produce flowers that are deformed or incompletely formed, incapable of pollination and therefore useless for producing fruit. In more severe cases, the plant gets confused about which direction to grow and produces branches that grow in many directions instead of predominantly one, looking like a "witches broom." In addition, boron deficiency alters the production of lignins and compounds related to them. This reduces the strength of the wood and may make the oil more bitter.[24]

Like many crucial nutrients in humans, too much boron is just as bad as too little. The difference between the correct dose and the overdose is quite small, so boron must be applied carefully and at the right time. It can be applied to the leaves or through the soil.[3]

3.2.2.3 Alternate Bearing. A curious and not fully understood characteristic of the olive tree is its tendency to produce a larger than average crop one year, then a smaller than average crop the following year (see Figure 3.10 for the effect of this on oil

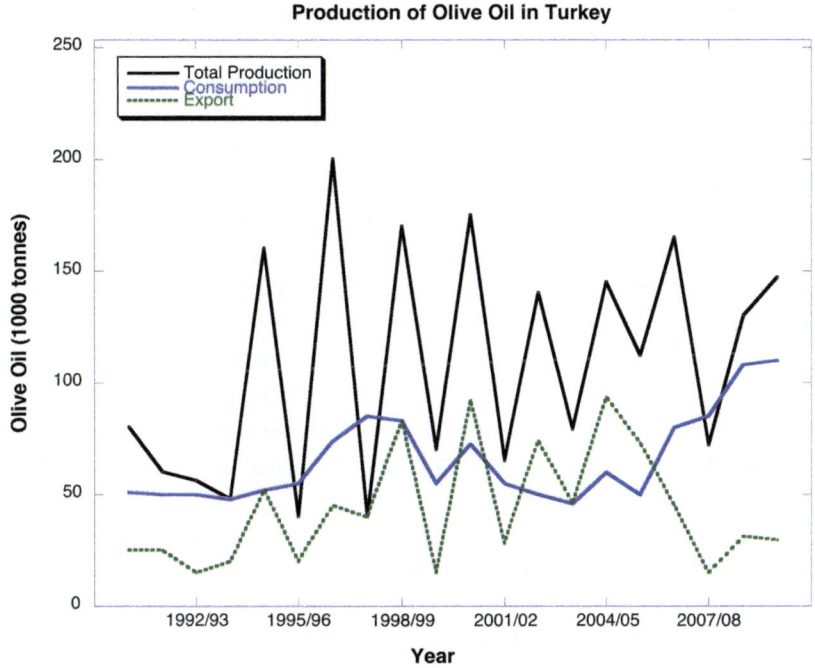

Figure 3.10 Annual production of olive oil in Turkey. Note the strong cycle of alternate bearing. Data from IOC country profile for Turkey.

production). This tendency, called "alternate bearing," is not limited to olives but is common among many fruit trees. The tendency arises from the tree's need to divide resources between enlarging the tree and producing offspring. Various hormonal signals in the tree can cause the division to swing from one goal to another from year to year. A bud on new growth wood is set on the path to becoming a flower or a leaf in the year before it actually develops. If the tree is heavily fruited in the earlier year, a signal directs the emerging bud to develop leaf attributes rather than fruit attributes. If, instead, the fruit yield is low (or the fruit harvested early), the emerging bud is more likely to develop into a flower in a process known as induction.[25] This will definitely affect alternate bearing and is a reason for the push to harvest fruit as early as possible in large groves. The tendency to alternate bearing is especially pronounced in trees with limited resources (such as nutrients and water).

In areas in which trees are not as actively managed, such as some in the Mediterranean, this alternate bearing is simply accepted as part of the process. It is quite a bit more difficult to harvest olives during an off year, as there are fewer olives per tree. However, if one owns more trees than can be easily harvested directly, some extra effort can make up for the lower fruit density. In more managed groves, especially in less traditional areas, there is active research into methods to minimize the difference in production between the off and the on years.

Such research must start with trying to understand the origin of the phenomenon. While the production of hormones and the size of energy stores in the tree make sense as a possible cause for alternate bearing, they were found to be unrelated.[26,27] Genetic analysis shows that carbohydrate metabolism is very different between the on and off year trees,[28] making a tantalizing suggestion for the cause, but begging the question about what causes the metabolism to change.

Most producers who manage their trees actively use a combination of relatively young trees, careful pruning to preserve flowering growth, nutrient delivery, irrigation, and early harvest to minimize the alternate bearing cycle.

3.2.2.4 Pollination. Olive trees pollinate by a wind-driven process, which means that the pollen drifts from one tree to the next and is captured by the flowers. For more intense agricultural farming, groves need to be designed with pollination requirements in mind. Unlike many other fruits (for example apples), many olive cultivars self-pollinate. However, fruit yield can be significantly increased by cross pollination. If no fruit is obtained in the absence of cross pollination, such cultivars are called "self-incompatible." Other than the right kind of self-pollinating cultivar, the environmental conditions such as wind direction and strength as well as the design of the orchard are also important factors for proper pollination.[29]

3.2.2.5 Springtime Insect Pests. During our exploration in South Africa, we encountered an unusual disease caused by Olive Seed Wasps.[30] This wasp lays its eggs in the tender early seed, before the flesh is thicker and the seed has hardened. The eggs hatch late in the season, and the larvae bore their way through the pit, and out of the flesh, leaving a hole in the fruit that causes spoiling. These pests are tricky, as there is no easily visible damage when the olive is attacked. It is only quite a bit later that the damage to the fruit is noticed.

3.2.3 Summer

The summer is hot and dry in Mediterranean climates, and the sun is at its highest, the days at their longest. This is a time of intense growth on the part of both the tree and the olives. After having survived for many millennia in a harsh environment, the trees have a number of tricks for dealing with the heat and growth, resulting in a very productive plant if treated well.

As a tree is optimized for growing in hot, dry climates, there are several strategies for extracting and conserving water available to the olive. One of the secrets is in the leaves. The silvery leaf does not absorb as much light as a darker one, a reflection on the abundant supply of sunlight and the dangers of overheating the leaves. In addition, the plant can close its pores very tightly during the day, reducing the loss of water from the leaves.

The other secret is in the roots. In areas of abundant water, the roots contain channels made up from cellulose (the basic building substance of wood). Cellulose does not dissolve in water, but it absorbs it very strongly. A paper straw can conduct water through it, but it does make it a bit harder to draw the water through. Olive trees, like other dry-weather plants, modify the channels with a different substance called lignin. This is the same material that is added to wood in stress points, as it makes the wood a bit harder and resistant to breaking. In the roots, the function is to line the root channels with a more hydrophobic substance, like the wax that is on most commercial paper straws, or the plastic that makes up others. This lining makes it easier for the water to flow through the channels. Interestingly, the components of lignin contribute to some of the bitter compounds in olive oil, so a tree grown under greater water stress will likely produce more bitter oil.

3.2.3.1 Irrigation. Like any plant, olive trees need water. Their needs depend on the growing conditions, in part – a dry, hot climate is the usual environment for the tree and causes the tree to lose a fair amount of moisture despite its best efforts to conserve. The minimum annual requirement for the tree is about 2–4 ML per hectare, at which point it will be alive, but not thriving (L. Ravetti, personal communication). Translating that into a more familiar description, the water needs would be satisfied by rainfall of about 20 cm, or a little less than eight inches, per year if it were completely captured by the tree. Even twice that amount matches the rainfall in desert regions. You can see that the olive tree is an admirably efficient tree regarding water consumption.

However, the fruit production on a tree with this minimal rainfall will be poor. Many olive growing regions have substantially greater rainfall, satisfying the need for more water (optimally about 6–8 ML ha^{-1}). This allows the tree to put energy into producing flowers and fruit. In addition, water can be added by a variety of irrigation strategies.

The simplest irrigation strategy is a periodic flooding of the trees. A moat can be made around a tree and water poured in periodically (perhaps once a month). In one case, we saw a farmer whose field was quite flat and near to an irrigation canal. He flooded the entire field once per season and allowed the water to sink in to the deep layers of soil below. This strategy works best when the soils are somewhat clay-like, allowing them to hold on to the water. A sandy, quick draining soil would not hold the water. We also watched a Turkish grower using a small pond built around each tree and filling it once per month (shown in Figure 3.11). This

Figure 3.11   Irrigation pond and olive tree in Edincik, Turkey. The sight of the Aegean Sea in the background reminds us of the saying that "olive trees like to be able to see the sea."

strategy has the advantage of being relatively simple, although it is likely not as efficient in the use of water as others.

Most of the modern farms have an irrigation system based on a simple-sounding drip technique. Pipes and tubing are run into the groves, and outlets are provided at a fixed distance from each tree trunk, allowing water to be pumped to each tree and delivered to the ground above the roots. Due to the fine control available in this system, it can work well in all soil types and can be tuned to deal with differences between one part of the grove and another. Depending on the size of the system, this can be fed by a tank of water and a simple pump turned on and off by the farmer, or by several reservoirs of water fed into a complex pump house with many pumps and a computer controlled distribution system. Given the critical nature of water and nutrients to the health of the tree, an enormous amount of thought and planning is given to this part of the production.

The timing of irrigation is a bit of a controversial topic. One could water frequently, but lightly, or infrequently, but heavily. Arend Hofmeyr, a grove owner and processor in South Africa brought two experts to view his land, Portion 36 near Stellenbosch. Standing in exactly the same place, within a week of each other, the two experts gave exactly contradictory advice. Most of the farmers have found that advice about irrigation is a lot like advice about child rearing. It is very useful to frame the issues and to help the farmer know what questions to ask. In the end, it is experience with the trees, watching them grow, and getting a feeling for what these trees in this place need that will answer the question of how to irrigate.

As mentioned above, drip irrigation allows the delivery of nutrients by mixing in water soluble nutrients for delivery at the right time. Nutrient needs will change over the season, making this near-continous provision of nutrients and water helpful.

3.2.3.2 Summer Pests. The Olive Fly (*Bactrocera oleae*) is a widespread pest, which lays eggs in the growing olive. When these eggs hatch, the larvae eat channels in the olive, which then cause the olive to spoil, as shown in Figure 3.12. If widespread, this can cause the entire harvest to produce poor-quality oil,

Figure 3.12   A fly-bitten olive in the hand of Josep Ramon Morera, in Catalonia.

with bad test results and a "fly-bitten" flavor. Treatment of the fly can have the effect of helping to protect against black scale as it enhances growth of natural predators. Many farmers use simple attractant baited traps to look for flies before spraying to reduce the amount of insecticide that must be used.

3.2.4   Fall

The fall is a very active season, culminating in harvest. The olives are growing rapidly, and the changes in the pigments during this season give a wonderful new look to the fruit. One of the most relevant changes is that the oil begins to accumulate in the olive. The spring is devoted to making the olive happen, the summer is devoted to making the apparatus for oil production in the olive, and the fall is when oil production is at its highest. The benefit to the tree is that an oily olive will finally look like something attractive to eat, and the animals that eat the olives will distribute the seeds widely after eating the fruit. Our benefit, of course, derives directly from the oil in the fruit, which reaches its maximum concentration late in the fall. Because fall is so intimately tied with the harvest, we will defer full discussion of this season until the next chapter.

3.3 THE BITTER END: WHERE DOES OLEUROPEIN COME FROM?

Plants, including olive trees, are complex natural collections of chemicals that interact in wonderful and productive ways. The cellulose that makes up the woody parts of the tree, the proteins that make up the cellular machinery, and the triglycerides that make up the oil are all formed in large amounts. There are chemicals that are present only temporarily to serve some function, such as the sugars used as intermediates to make other chemicals. There are even chemicals whose job is to make other chemicals, such as the enzymes that guide and accelerate the reactions making the oil in the olives, or to make the molecules responsible for the beautiful purple color of a ripe olive. These chemicals are all made by chemical reactions, guided by the enzymes in the plant, which in turn are produced by reading the plant's DNA. These chemicals are the expression of what it means to be an olive tree.

Practically minded people will be reassured that this is not magic. The raw materials for the molecules must come from the environment, and if there is a shortage of some raw materials, less of the chemicals will be made. The tree uses material from the water in the soil (water itself, minerals, and nitrogen-containing nutrients, among others). It also uses material from the air (carbon dioxide and oxygen primarily). And, of course, it uses sunlight as a source of energy. This energy is required because many of the reactions are energetically difficult.

We will tell this story about oleuropein, a compound that makes olives taste very bitter, helps protect against predators, and contributes to an important family of compounds in olive oil responsible for key aspects of flavor and many health benefits. Oleuropein is a wonderful example of a complex molecule made by a natural process. The story starts with the raw materials and discusses how these are converted into the subunits of the molecule, which are assembled to make the final product.

3.3.1 Key Starting Materials

A plant growing in a harsh environment needs to balance the need for complexity of chemicals to carry out the many functions required with the need for efficiency in the production of these

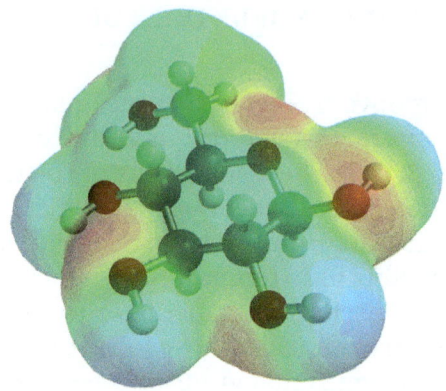

Figure 3.13   Structure of glucose.

chemicals. One solution to this seeming paradox is to use some key intermediate chemicals to serve in the production of lots of different final products. These intermediates can be slightly more complex than the raw materials, but not so complex that their ultimate use is limited.

One of the key starting materials that may be familiar to you is glucose, as shown in Figure 3.13. This is a simple six-carbon sugar that is easily digested by humans as well as olive trees and is the primary product of photosynthesis. It is also the primary building block of cellulose, which makes up the wood of the olive tree as well as some other structural parts. Since it's so important, it is produced in very large quantities by the plant, and there is almost always a lot on hand.

Another important starting material is an intermediate breakdown product of glucose called acetyl coenzyme A, or acetyl CoA for short. Acetyl CoA, shown in Figure 3.14, is used by the plant to make many different products, including the fatty acid molecules, some of the pigments, and the rare molecules that are produced to tell insects to go away before the olives are ready. The molecule delivers the two-carbon acetyl fragment to a molecule being built, and recycles the large and complex CoA fragment to be used with another acetyl group. We'll meet acetyl CoA again in Chapter 8.

The third key starting material is an amino acid called phenylalanine, shown in Figure 3.15. Amino acids are the building blocks of proteins, and are present in all living tissue. Of the 20

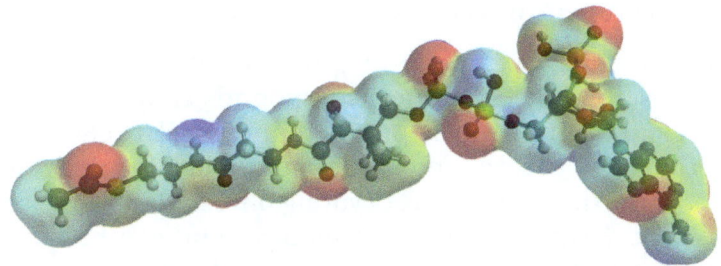

Figure 3.14 Structure of acetyl CoA. The portion used to build molecules is the two-carbon tip at the very left end. The remainder of the molecule is a recyclable handle to ensure it gets to the correct location in the cell.

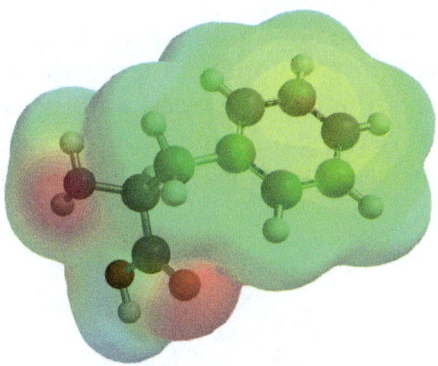

Figure 3.15 Structure of phenylalanine. Note the aromatic ring on the right.

amino acids, phenylalanine is one of the larger ones and contains one of the structural features of antioxidants, a flat array of carbon atoms called an aromatic ring.

From these three key chemicals, glucose, acetyl CoA, and phenylalanine, almost all of the components of olive oil can be produced by the plant. This makes for a very efficient and responsive production. This is true both for chemicals which are needed in large amounts and for those needed in tiny amounts. The latter are sometimes called secondary metabolites and serve various functions, such as giving color to the olive fruit in fall, signaling stages of development, and preventing insects from eating the olive, to name just a few. The secondary metabolites are represented here by oleuropein, which is found uniquely in the olive and its cousins.

3.3.2   Oleuropein, a Secondary Metabolite

Oleuropein ("ole-you-rope-ee-in"), one of the more well-known and intensely studied chemical compounds associated with the olive, is shown in Figure 3.16. It is found in relatively high concentrations in the leaves and in the young fruit, and can reach up to 14% of the dry weight of very young olives.[31] Its concentration in the fruit declines as the fruit matures and can be quite low in very ripe fruit. Its very bitter taste can deter grazing animals and birds tasting the unripe fruit, and it can also prevent some insects from feeding on the olives or leaves by making them indigestible.[32] Defense against predators is a common use for secondary metabolites and is very important for a tree, which cannot run away or swat flies.

Interestingly, oleuropein is found in a wide variety of the olive's cousins, such as European and Chinese ash trees, several lilac varieties, pink or white jasmine trees, and the fragrant olive. Of these trees, the olive is the only one to produce edible fruit (while the flowers of the fragrant olive are widely used for tea and jam, the fruit is not).

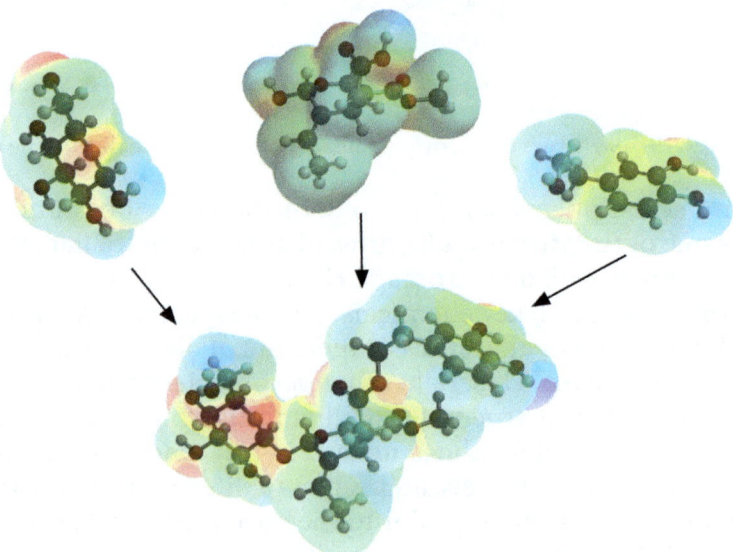

Figure 3.16   Structure of oleuropein (bottom). This molecule is made from glucose (upper left), elenoic acid (top), and hydroxytyrosol (upper right).

Those who have tasted an unripe, uncured olive have met oleuropein in high doses. While very vivid, the experience is not entirely pleasant; a bit like meeting a famous person who insults your family. It leaves a long-lasting bitter taste in the mouth. In smaller doses, with quite a bit of sugar, it can be enjoyable, as in the form of olive leaf tea. The compound has a wide variety of beneficial health effects that will be described later in this book.

From a chemical standpoint, oleuropein (shown in Figure 3.16) is a molecule that is grafted together from three sources. The molecule is built from a water-soluble glucose molecule grafted onto a non-polar structural core (elenolic acid) which in turn is connected to a hydroxytyrosol molecule that enhances bitterness and gives antioxidant activity.

Of these, glucose is the most familiar molecule. As stated above, it is easily available in all plants. It also has a lot of chemical "hooks" that allow it to be easily attached to other molecules. Hydroxytyrosol, shown in the top right of Figure 3.16, is not well-known outside the circle of experts but is available to almost all woody plants. It has its origin in the chemical production pathways that yield lignans, complex polymers very common in woody tissues of trees. In some trees, lignans are second in quantity only to cellulose and convey strength to woody tissue or, as discussed above, enhance the efficiency of the water channels in the roots. One thing to note here is that scaling up the production of lignans adds to the supply of hydroxytyrosol and molecules related to it, and may enhance the supply of oleuropein.

The core of the compound is the relatively complex molecule called elenolic acid, shown in the top center of Figure 3.16. This is the part of the process that is unique to olives and their cousins. Starting from acetyl CoA, a long chain of chemical reactions is managed by the plant enzymes to combine six acetyl CoA molecules, shave off a couple of carbon and oxygen atoms, link parts of the molecule together to make two rings, add oxygen atoms in different locations, and clip open one of the rings to produce a reactive location for grafting. This may seem terribly complex, but it is very similar to industrial processes that make a coffee pot, windshield wiper, or flashlight. Each step is performed by an expert optimized to do that one job with maximum efficiency, although in this case the "experts" are enzymes. We will say more about enzymes in Chapters 5 and 8.

Elenolic acid is not very useful by itself, but with its glucose molecule grafted in one place and the hydroxytyrosol grafted in the other (separate enzymes specialize in these two steps to get each in the correct spot), it makes oleuropein, the powerful anti-oxidant and anti-feedant.

We are focusing on oleuropein because it is the primary source of the array of bitter compounds in olive oil. Even though it does not appear in the oil, due to its high polarity and affinity for water, enzymes that can remove the glucose portion release less polar molecules that do appear. These enzymes are usually kept separate from the oleuropein in the olive, preserving the extreme bitterness. There is one exception that we have encountered: the Hurma olives in Turkey and similar locations.

3.3.3   Naturally Debittered Hurma Olives

Along the Aegean Coast, on the Karaburun peninsula of Turkey, a most miraculous collision of botany and microbiology has produced an olive known as the Hurma olive that is naturally debittered. In Turkish, "hurma" means the date fruit, and indicates the sweet nature of the Hurma olives when eaten directly off the tree. Olives picked from these groves, shown in Figure 3.17, can be eaten immediately and are as delicious as others soaked and brined for weeks to months to remove the bitter phenolic compounds. While these olives seem to be identical in variety to thin rinded Erkence black olives, the Erkence show a normal high phenolic content while the Hurma show a significantly lowered phenolic profile.[33] The position of the Karaburun peninsula in the prevailing winds creates a microenvironment with a constant supply of moist breezes from the north. It has been suggested that a local fungus with a very small indigenous area is carried on the sea breezes to transform the Erkence olives into the Hurma by converting the phenolics to sweeter secondary metabolites through the action of the bacteria or fungus. Microbiological analysis of the Throuba Thassos olive, a similar naturally debittered olive from Greece has identified the fungus *Phomo olea* that breaks down the oleoeuropein in the mature fruit on the tree, resulting in a sweetened olive. Other studies report a similar olive, the Dhokar, from Tunisia.[33] Having olives that are naturally cured would be useful to those hypertensive individuals who must restrict salt intake. The longer shelf life of the Hurma olive

Figure 3.17 Naturally debittered Hurma olives in Turkey can be eaten right off the tree.

fruit (40 days as compared with three to four days for a freshly picked olive) makes them highly prized by their fans.

3.4 WHAT'S NEXT?

As you saw in Chapter 2 and in this one, the olive is busy throughout the growing season making triglycerides and phenolics such as oleuropein. These are only two of the essential chemicals that are required for high-quality olive oil. At the end of the season, however, the production of these compounds reaches its peak levels. Analysis of these and other species, along with the visible clues such as the color of the olives, tells us that it's time to reap the benefit of all this hard work. It's harvest time.

REFERENCES

1. International Olive Oil Council, *World Olive Encyclopedia*, World Olive Council, Madrid, Spain, 1996.
2. *Olive – Olive Growing, Olive Oil and Table Olives*, http://www.fosfa.org/wp-content/uploads/2014/11/Olive.pdf, accessed May 2015.

3. X. Rius García and J. M. Lacarte Peña, *The Olive Growing Revolution : The Super High Density System*, Waikerie, Australia, 2012.
4. X. Arnan, B. C. Lopez, J. Martínez-Vilalta, M. Estorach and R. Poyatos, *Dendrochronologia*, 2012, **30**, 11.
5. L. J. Fick, G. H. Fick, Z. Li, E. Cao, B. Bao, D. Heffelfinger, H. G. Parker, E. A. Ostrander and K. Riabowol, *Cell Rep.*, 2012, **2**, 1530.
6. A. Fabbri, G. Bartolini, M. Lambardi, and S. Kailis, *Olive Propagation Manual*, Landlinks Press, Collingwood, VIC, Australia, 2004.
7. R. Efe, A. Soykan, İ. Ciberal and S. Sönmez, *Dünyada, Türkiye'de, Edremit Körfezi Çevresinde Zeytin ve Zeytinyağı*, Edremit Belediyesi, Balıkesir, 2013.
8. C. Conde, S. Delrot and H. Gerós, *J. Plant Physiol.*, 2008, **165**, 1545.
9. H. B. Ash, *On agricultura. I: Res Rustica*, 1977.
10. *How Copper Sprays Work and Avoiding Phytotoxicity – Cornell Vegetable Program – Cornell University – Cornell Cooperative Extension*, http://cvp.cce.cornell.edu/submission.php?id=140, accessed February 2016.
11. C. Ramos, I. M. Matas, L. Bardaji, I. M. Aragón and J. Murillo, *Mol. Plant Pathol.*, 2012, **13**, 998.
12. F. J. López-Escudero and J. Mercado-Blanco, *Plant Soil*, 2011, **344**, 1.
13. P. Vossen, D. Gubler and M. A. Blanco, *Newsletter of Olive Oil Production and Evaluation. Univ. of California Cooperative Extension*, 2008, vol. 3, p. 1.
14. T. H. Thomas, *Plant Growth Regul.*, 1998, **26**, 141.
15. Y. Ben-Dov and C. J. Hodgson, *Soft Scale Insects: Their Biology, Natural Enemies and Control*, Elsevier Science Publishers, 1997, vol. 7a.
16. *Olive Tree Felling Starts in Apulia Amid Protests*, http://www.oliveoiltimes.com/olive-oil-making-and-milling/olive-tree-felling-starts-in-apulia-amid-protests/47326, accessed February 2016.
17. *Xylella fastidiosa – European Commission*, http://ec.europa.eu/food/plant/plant_health_biosecurity/legislation/emergency_measures/xylella-fastidiosa/index_en.htm, accessed February 2016.

18. A. Abbott, *Nature*, 2015, **522**, 13.
19. *Nutritional Recommendations for Olives*, http://www.haifa-group.com/files/Guides/Olive_Booklet.pdf, accessed February 2016.
20. F. López-Granados, M. Jurado-Expósito, S. Alamo and L. Garcia-Torres, *Eur. J. Agron.*, 2004, **21**, 209.
21. R. Fernandez-Escobar, A. Ortiz-Urquiza, M. Prado and H. F. Rapoport, *Environ. Exp. Bot.*, 2008, **64**, 113.
22. R. Erel, U. Yermiyahu, J. Van Opstal, A. Ben-Gal, A. Schwartz and A. Dag, *Sci. Hortic.*, 2013, **159**, 8.
23. K. Noguchi, T. Ishii, T. Matsunaga, K. Kakegawa, H. Hayashi and T. Fujiwara, *J. Plant Nutr. Soil Sci.*, 2003, **166**, 175.
24. N. Wang, C. Yang, Z. Pan, Y. Liu and S.-A. Peng, *Front. Plant Sci.*, 2015, **6**, 916.
25. D. J. Connor and E. Fereres, *Hortic. Rev.*, 2005, **31**, 155.
26. S. M. Al-Shdiefat and M. M. Qrunfleh, *Jordan J. Agric. Sci.*, 2010, **4**, 12.
27. A. Bustan, A. Avni, S. Lavee, I. Zipori, Y. Yeselson, A. A. Schaffer, J. Riov and A. Dag, *Tree Physiol.*, 2011, **31**, 519.
28. M. Turktas, B. Inal, S. Okay, E. G. Erkilic, E. Dundar, P. Hernandez, G. Dorado and T. Unver, *PLoS One*, 2013, **8**, e59876.
29. G. C. Koubouris, C. M. Breton, I. T. Metzidakis and M. D. Vasilakakis, *Sci. Hortic.*, 2014, **176**, 91.
30. P. Neuenschwander, *Z. Angew. Entomol.*, 1982, **94**, 509.
31. V. Goulas, P. Charisiadis, I. P. Gerothanassis and G. A. Manganaris, *Curr. Bioact. Compd.*, 2012, **8**, 232.
32. K. Konno, C. Hirayama, H. Yasui and M. Nakamura, *Proc. Natl. Acad. Sci. U. S. A.*, 1999, **96**, 9159.
33. A. B. Aktas, B. Ozen, F. Tokatli and I. Sen, *J. Sci. Food Agric.*, 2014, **94**, 691.

CHAPTER 4

Season's End: Harvesting the Fruit

The growing season is over, the olive branches are laden with fruit, most of the other crops have already been brought in, and there's a chill in the air. The pruning, irrigation, fertilization, and pest management are all over. In many parts of the world, this means it is time to harvest olives. This time of year is a culmination of all of the hard work described in the previous chapters.

Growers have three goals in mind when they decide about harvest. One goal is, of course, maximizing quality. The second two goals are more economic: maximizing quantity and minimizing cost. In classic fashion, these goals cannot all be maximized at the same time.

Quantity and quality of oil are largely determined by the time of harvest, as we will elaborate below. However, the timing of the harvest is more critical for olives than many other fruits. Olives need to ripen on the tree, unlike other fruits that ripen after harvest (called climacteric fruit). They also do not release or respond to the ripening gas ethylene in the same way as other fruits such as bananas or avocados.[1] Therefore, the olives need to be picked at the perfect moment for maximum quality.

Harvest time also affects yield of oil. If growers pick too early, the yield of oil per quantity of olives might be too low. Oil content typically increases through the season, and oil that has not yet

The Chemical Story of Olive Oil: From Grove to Table
By Richard Blatchly, Zeynep Delen Nircan and Patricia O'Hara
© Richard Blatchly, Zeynep Delen Nircan and Patricia O'Hara, 2017
Published by the Royal Society of Chemistry, www.rsc.org

developed in the olive can, of course, not be extracted. On the other hand, if the olives are picked too late, many of the olives might fall naturally to the ground, reducing the number of olives on the tree. It is easy to be distracted by a fat yield of olive oil in your batch of harvested olives and not realize that you harvested only a fraction of the number on the tree earlier in the season.

Fallen olives can easily become infected or mildewed or bruised, and cannot be used for virgin or extra virgin oil or for table olives. They can however, be pressed and the poor quality oil that results be transformed into refined oil in a factory. Here, the focus will be on olives harvested to make high-quality olive oil.

Farmers must also choose the best method to pick the olives. This choice is complex, depending on the size of the grove, the local labor market, the capacity of the nearby processing facilities, the profile of the grove, and many other factors. There can be remarkable uniformity of harvest methods in a region, and remarkable diversity.

4.1   WHEN ARE OLIVES HARVESTED?

Like any fruit, olives respond to environmental cues such as lower temperatures and fewer hours of daylight to signal the end of the growing season. As the fruit ripens, a scar forms between the stem and the olive to ready the fruit to drop to the ground in a process known as abscission. If farmers pick too early, this scar has not yet formed and the force necessary to pick the fruit off the tree can damage the fruit or the tree. Once the abscission scar is complete, the fruit can be easily picked or can in fact fall off the tree in a light wind or when the tree is shaken. Since the ultimate biological purpose of ripening is for the fruit to be eaten by animals so the seeds (olive pit) can be dispersed, this makes sense.

By harvest time, a single olive tree can produce up to 80 kg of olives or 60 000 olives! Of course, the yield for any particular tree will depend on cultivar, size of the tree, management of the grove, and the luck of the season.[2,3] As the olives mature on the trees, they also become enriched with oil. A young green olive might contain 5% oil by mass while a ripe black olive can contain up to 35% oil. The numbers reported here are for maximum production yields;

it is most typical for traditional farmers in the Mediterranean to report yields that average 8–10 kg of olives per tree and, with a 15% yield, this translates into 1–2 liters of oil per tree.[4]

4.2   WHO DECIDES WHEN TO HARVEST?

Due to the critical nature of the harvest timing, there will be one person of great responsibility who decides when to harvest. In the most modern groves, a technical manager will be making measurements on the growing olives through the season, developing a baseline from which the harvest decision can be made. This decision would be made in consultation with the managers of production and staff. In a smaller grove, the owner or manager will decide, based on a variety of factors including the color of the olives and perhaps some technical measurements. In the smallest groves, the decision is up to the owner, of course, but is often done in coordination with a local expert, who looks more holistically at the olives, observing the color, testing the oiliness of an olive broken open, or feeling the softness. One of our students in Turkey said, simply, "Oh, my grandmother just knows when to harvest. She goes into the grove every day, and says one day: It's time."

Harvest decisions are most often made on the basis of the color of the olives, an easy quality to observe. However, oil content is also important and can be measured casually (by breaking open the olive) or carefully (by using a near-infrared spectrometer).

Most table olives are picked on a different schedule – earlier harvest for the green table olives and late harvest for the Kalamata, Mission, and many other black table olives that are naturally darkened. For processed black olives, ferrous gluconate can be added to artificially darken and stabilize the color. If a cured table olive still contains a pit, one can see if the olive has been artificially darkened as the pit will be blackened as well. In a naturally cured olive, the pit will remain brown.

4.2.1   What Does the Color of an Olive Tell You About Ripeness?

Of course, size of the olives and the oil content are important, but a farmer will also carefully keep track of the color of the olives. As the season progresses, the olives will begin to lose their green color and, depending on the cultivar and the terroir, gradually turn an intermediate color that can be either speckled purple, half green/half purple, or a creamy pink/purple before turning dark purple or black. This intermediate color is given its own name, veraison, meaning "the onset of ripening" and is the same word as that used to describe the color at which wine grapes begin to ripen. Virgin oil made from fruit at this stage will taste fruity and have varying amounts of bitterness and pungency. A bit more time on the trees and most olive fruit will turn black. Some consumers prefer the taste of oil made from a later harvested blacker olive, which will be less bitter (and so will taste sweeter) and the yield of oil per olive is usually higher. Figure 4.1 shows the size and color progression of olives as they ripen and the colors of the oils they produce at various stages. Figure 4.2 shows the olives still on the trees in different degrees of ripeness.

4.2.2   Where Does the Color Come From?

What exactly is happening in the fruit to change the color from green to black? The green color of immature fruit is due to members of the chlorophyll family, the light-harvesting molecule present in all green plants (see Chapter 2). With photosynthesis no longer important to the fruit, the chlorophyll molecules are

Figure 4.1   Size and color changes of olives and oil from IOC November 2011 Characteristics of Oil-Olives, photograph Jos. Alba Mendoza.

(a)   (b)

(c)   (d)

Figure 4.2   Olives on the tree in varying degrees of ripeness: (a) immature
green Leccino olives in South Africa; (b) pink Ayvalik Olives in
Assos, Turkey; (c) half green, half purple Kolovi olives in Lesvos,
Greece; (d) pink speckled Koroneiki from Kritsa, Crete (photos
Rich Blatchly and Zeynep Delen).

broken down and the green color fades.[5] As the green fades, the fruit begins to make a new class of molecules, unrelated to the chlorophyll family, called anthocyanins. Anthocyanins, which are deeply colored, are extremely interesting on their own – they can act like sunscreen to protect a plant from sun damage – but they are also meant to attract birds to eat the fruit so that seeds will be dispersed. The water-soluble anthocyanins in ripe olives are similar to those in beets, eggplants, plums, and other deeply colored fruit, vegetables, and flowers and are also a great dietary source of antioxidants.[6] Figure 4.3 shows a plot of the absorbance of light for each wavelength by two different chlorophylls and an anthocyanin called malvedin (quite similar to the luteolin found in olives). Notice how the absorbance curve for the anthocyanin covers almost all of the wavelengths of visible light – letting just a bit of blue and red light through to produce the glorious purples of ripe olives, shown in Figure 4.2, and the color of the ripe olives in Figure 4.3(b).

Figure 4.4 shows one of the anthocyanin molecules responsible for this color. From the strong polarity of the molecule (see all the red and blue on the surface?), it is not hard to imagine

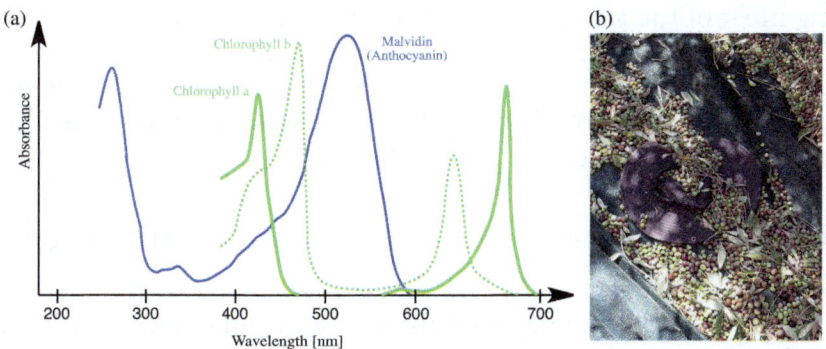

Figure 4.3 Development of color in olives: (a) absorption spectra of chlorophylls [green lines] and anthocyanins [purple line]. Chlorophyll absorbs purple and red light, so it appears green. Anthocyanins absorb light throughout most of the visible spectrum, passing a bit of blue and a bit of red, so they appear purple–black. Image adapted from http://en.wikipedia.org/wiki/Anthocyanin (b) Olives harvested from Morganster Estate, South Africa, showing various stages of ripeness.

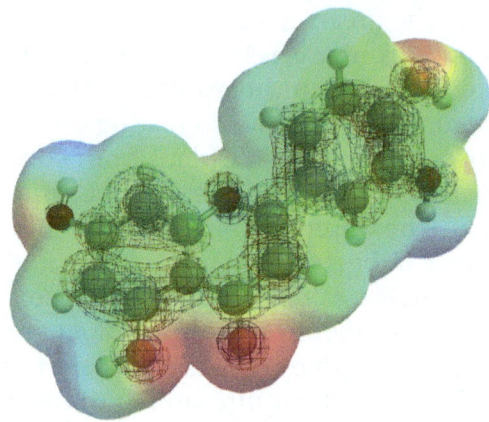

Figure 4.4   The anthocyanin luteolin, found in several forms in the skins of ripe olives. The increased presence of this molecule helps give the ripe fruit its characteristic purple color.

that this molecule is very water soluble, so will not appear in the oil when it is produced. Found in the skins of the olives, and serving as a sunscreen as the chlorophyll disappears, this molecule is excellent at absorbing light. The mesh that you see surrounding most of the atoms in the molecule is the molecular antenna responsible for interacting with light. In this case, rather than using that energy to generate chemical potential, the light energy is converted to thermal energy, warming the olive but not allowing the sunlight to cause chemical damage.

4.2.3   What Does Oil Content Tell Us About Harvest Timing?

Exactly how much oil does the fruit contain? 12%? 15%? 18%? Is the percentage going up? Holding steady? Going down? Answers to these questions are of utmost importance at harvest time. In the grove, farmers may sample the fruit frequently to determine the moment when the increase in oil begins to level off. Since most groves contain several cultivars, and it is likely that they will be ready to harvest at different times, each cultivar on the grove must be sampled separately. The percent of oil might determine how much a producer pays a farmer for his olives or how many liters of oil producers will be able to collect once the fruit is pressed. Later on, at the processing plant, regular measurements of percent of oil in the incoming fruit and comparison to the volume of oil produced will tell a plant manager whether or not the olive press is operating at a peak efficiency. Even small losses in the oil can translate into huge capital losses – imagine a facility processing 50 000 kilos of olives per day and producing only 9500 liters (19%) of extra virgin olive oil (EVOO) rather than the 10 000 liters (20%) that the olives contained. With the financial margins so tight in the olive production business, this inefficiency can turn a small profit into a loss when considered over the growing season.

To help track these important numbers, most modern producers will take advantage of a simple-to-use analyzer that measures the percent of oil in the olive paste or pomace by infrared spectroscopy.[7] A few olives are ground up; the paste is placed in the analyzer. Inside the instrument, infrared light is directed onto the sample (technically, this is called "near-infrared" or NIR, as it is very close in energy to visible light). Since the energy of this light matches the energy of a carbon–hydrogen bond stretch in the oil, some of the light is absorbed. The amount of oil can be related to the decrease in the intensity of the light. The analyzers are calibrated with known samples and the output read to a computer within minutes.

The rule of thumb that we heard from our visits was that harvest should be considered when the amount of oil in the olives stopped climbing and plateaued. While this sometimes correlated with the color changes described above, it could also be found in olives that were still completely green.

A routine NIR instrument, as shown in Figure 4.5, may cost $30 000, a very significant cost in this business. Large groves with very high throughput generally require them and can justify the expense much more easily than can smaller ones. For a significantly higher price, new more sophisticated instruments measure oil content, acidity, and moisture, even in intact olives on a conveyor belt, by near Infra-Red Diode Array Spectroscopy[8] or Fourier Transform Near Infrared Spectroscopy.[9]

4.2.4 Where Does the Oil Content Come From?

In Chapter 3, we introduced acetyl CoA, an important starting material in the making of oleuropein, an olive antioxidant in the leaves of the tree. Acetyl CoA is also used to make a wide variety of larger molecules. Of particular interest is its use in the production of fatty acids in the growing olives.

Figure 4.5   Olive paste in the tray of a near-infrared (NIR) machine awaiting analysis in Kritsa, Greece.

In Figure 4.6, we can see the repeated use of acetyl CoA to build fatty acids two carbons at a time. In the first process, two acetyl CoA molecules are combined to chemically stitch the end fragments into a four-carbon fatty acid. This process requires considerable energy input and an enzyme catalytic system, but in the presence of sufficient acetyl CoA and energy, is fairly efficient. The elegance of the process is that the same chemical process and catalyst can be used to add two more carbons for a total of six, and then again two more to make eight. The color codes show that each pair of adjacent carbons come from one acetyl CoA. This explains why the fatty acids almost always contain an even number of carbons. Ultimately, the process continues until the fatty acid of the required size is produced (the 18 carbon stearic acid is shown here).

If unsaturated fatty acids are needed, as they are for olives, the saturated molecules are converted to unsaturated by the action of enzymes that remove a pair of hydrogen atoms from a specific location, as shown in Figure 4.7. The first to be made is oleic acid, with a single unsaturation, then linoleic with two, then linolenic with three. Oleic acid is often referred to as a monounsaturated fatty acid (MUFA), and linoleic and linolenic acids are examples of polyunsaturated fatty acids (PUFA).

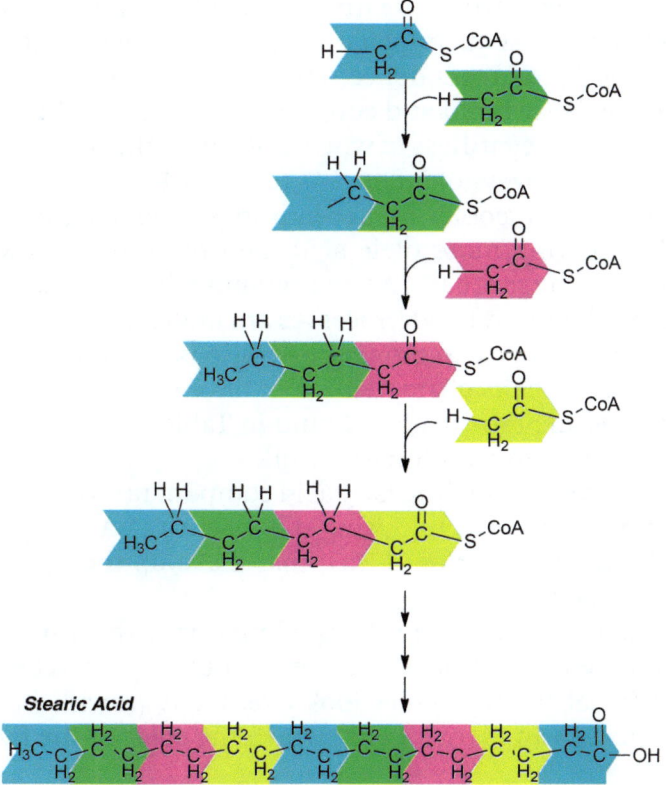

Figure 4.6   Synthesis of fatty acids, schematic form.

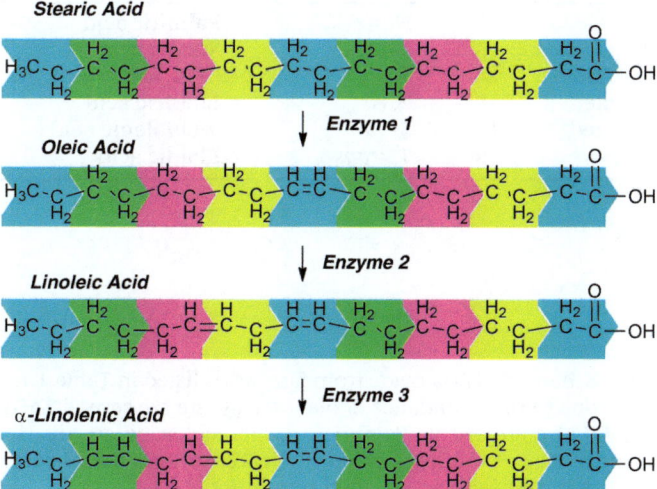

Figure 4.7   Conversion of saturated stearic fatty acids to a monounsaturated oleic fatty acid (MUFA), then to the polyunsaturated linoleic and linolenic fatty acids (PUFAs).

The particular mixture of fatty acids in olive oil is at least partly controlled by the sequential nature of these transformations and their temperature dependence. The conversion of stearic acid to oleic acid is fairly rapid and consistent, leaving very little stearic acid in olive oil regardless of when and where the harvest occurs. However, the conversion of MUFAs to the PUFAs is dependent on latitude, with the cooler higher latitudes producing less PUFAs (and thus leaving more oleic acid) and the warmer lower latitudes producing more PUFAs and retaining less oleic acid.[10]

These and other related processes are under genetic control of the olive, in response to growing conditions. Therefore, the mixture of fatty acids is characteristic of olive oil. A list of many of the contributing fatty acids is found in Table 4.1, along with the representative contribution to the oil.

After the synthesis of the fatty acid components, they are chemically attached to glycerol, using the reactive CoA end group, to make the triacylglycerides (TAG) that make up the oil, as introduced in Chapter 1.

Because the fatty acids are distributed into the TAGs according to their contribution to the overall mixture, it can be hard to visualize what the molecules look like. We've provided a simple schematic view in Figure 4.8, in which the color of the fatty acid

Table 4.1 Fatty acids found in olive oil.

Type	Size	Unsaturation	Common name	Appr.%
Saturated	16	None	Palmitic acid	10
Saturated	18	None	Stearic acid	2
Monounsaturated	18	1, *cis*	Oleic acid	70
Polyunsaturated	18	2, *cis*	Linoleic acid	10
Polyunsaturated	18	3, *χις*	α-Linolenic acid	1
Trans-fat	18	1, *trans*	Elaidic acid	0

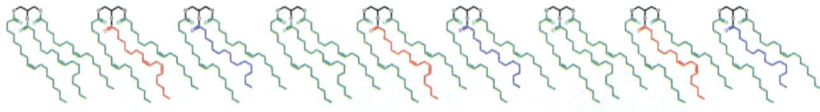

Figure 4.8 Schematic TAGs made from fatty acids listed in Table 4.1. Note that, due to the abundance of oleic acid, there are several TAGs with just that component. Palmitic and linoleic acids are distributed into about 10% of the positions each. These TAGs would break down to give the fatty acid profile shown in Table 4.1.

line drawing matches to colors given to the three most common fatty acids in Table 4.1.

4.2.5 Some Practical (and Impractical) Concerns that Affect Harvesting

Choosing exactly when to harvest olives is not all science. If a farmer must transport his olives to a community or cooperative press, practical concerns such as when the mill is available or when the laborers are ready may determine when the olives are picked. In one cooperative press we visited, picking was delayed for a day when the press broke down and a part had to be replaced. If equipment must be rented for a mechanical harvest, its availability will also impact the decision. Rain just prior to harvest is not good – the olives become swollen with water; and rain during harvest is also not good – the olives are more likely to mold if they are wet. So a farmer may choose to wait for clear, cool, dry weather or pick a bit earlier than he would like to take advantage of the proper weather conditions.

In many small towns and villages throughout the Mediterranean and the Middle East, the olive harvest is a collective and cooperative enterprise. It is unusual for one producer to be able to harvest all of his or her trees by themself. Harvest time might happen when it is possible to get friends and family or day laborers together.

Large scale growers with many hundreds of thousands or millions of trees have different concerns. For any particular cultivar, fruit maturation happens at the same time. To be harvested at the perfect moment, olives need to be picked during a short span of time: a few days or at most a few weeks. Figure 4.9 shows that at Boundary Bend farm in Australia, over 70% of their million trees are harvested in one month. By picking early, as is done at Boundary Bend, less fruit is lost to the ground. Other studies (results not shown) reveal a beneficial effect on the next year's crop when picking is done early in the maturation cycle.

To synchronously bring in all these olives is a challenge. To accommodate this, giant mechanical harvesters are used that can harvest 24 hours a day and in which one harvester can do the work of many humans. The largest machine, the MaqTec

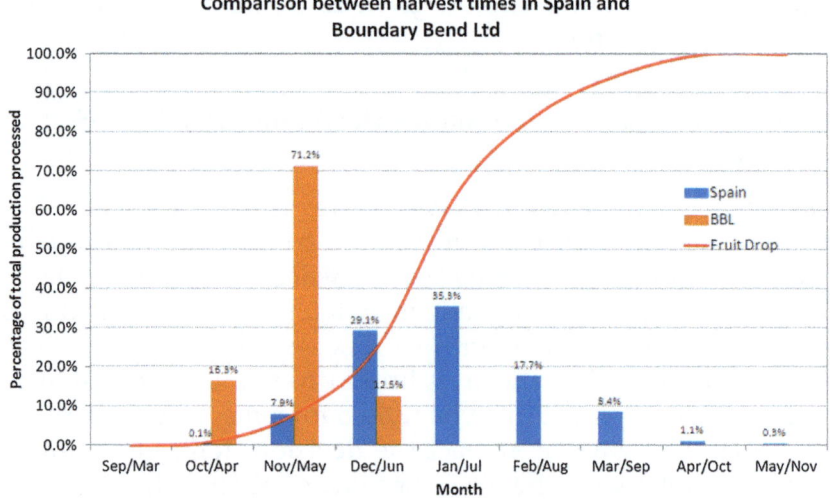

Figure 4.9   Comparison of harvesting schedule with fruit drop in two geo-graphic regions: Boundary Bend Farms in Australia, and Spain; information about harvest timing in Spain was obtained from AICA (http://www.aica.gob.es/). Permission Boundary Bend Ltd.

Figure 4.10   Colossus overhead harvester. Permission Boundary Bend Ltd.

Colossus shown in Figure 4.10, can harvest about 100 trees per hour. With a grove of one million trees, one Colossus work-ing non-stop would take a minimum of 40 days to harvest the grove. In reality, it takes much longer than this because the har-vesters need to turn around at the end of each row (not a trivial

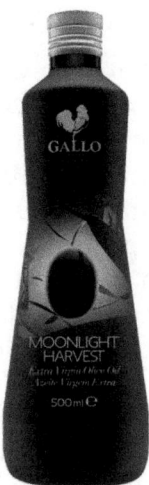

Figure 4.11 Moonlight Harvest Oil, produced by Gallo Olive Oil in Portugal. This oil is "born from moonlight harvesting, when the coldness of the night preserves all the fruit's aromas and flavors" and is reported to be "fresh, fruity and slightly spicy." Used with permission from Gallo Olive Oil, Photo credit Ana Calado.

task for such a large vehicle), and must stop for service and shift changes. If the goal is to create the highest quality oil, it is also necessary to have the fruit harvested when it is just ripe. As mentioned, while the intermediate color – veraison –is one indication, the percent oil is what most large groves rely upon. Keeping the olives on the tree after they are ripe may cause the oil produced to lose value as health giving compounds that contribute to the bitterness are lost through fermentation.[11] It is much more likely that groves with more than a few hundred thousand trees will have multiple mechanical harvesters. Still, the task of collecting the fruit from a million trees and getting it to press quickly is daunting.

A few boutique manufacturers advertise exquisite results when producing oil from olives picked only at night or under moonlight (Figure 4.11). As the promotional literature claims, the lower temperatures of evening and perhaps even the different night cycle of plants might explain some of the higher quality. Since olives decay more slowly when kept cold, night harvested olives could be superior to those picked by daylight simply because it is cooler at night.

4.3 HOW ARE THE OLIVES HARVESTED?

Harvest is about to start, but how will the olives be collected? The manner in which the olives are collected is one of the key features in determining the quality of the olive and the cost of production. Slow or careless picking can damage both the olives for this year's crop, producing low quality oil, as well as the trees and buds for next year's crops.

One major consideration is time. Olives need to get from the tree to the press as rapidly as possible to avoid simple degradation of the natural fruit that can produce an off-flavor and to minimize contamination from such culprits as mold that can grow easily on fruit held too long in burlap sacks or from soil bacteria from olives kept on the ground too long before being transferred to sacks. Similarly, bruised fruit decays even more rapidly than a carefully picked fruit. A bruise is a rupture in the plant cell wall that releases enzymes that break down the remaining fruit.

Beyond these very important concerns, the harvest method that makes most sense depends on the type of grove that has been planted. Figure 4.12 shows groves planted at different densities of trees. Traditional low density groves with beautifully tall old trees spaced far apart on terraced hillsides cannot be easily harvested by mechanical harvesters and are often picked by hand. At the same time, modern high density or super high density groves with 100 000 or more trees would take a prohibitive amount of time to harvest by hand.

4.3.1 Hand Harvesting

Long before we knew how to write, before we even knew how to weave cloth, humans knew how to harvest olives simply by picking them or causing them to fall with a stick. Hand-picking is quite tedious, labor intensive, and yet with trained pickers, produces high quality fruit. To help somewhat, small hand-held plastic rakes are often used to speed the process (Figure 4.13).

When trees grow too tall for simple picking, long sticks are used to shake the branches and the olives collect into nets on the ground. While this sounds simple, there are several do's and don'ts when it comes to hitting a tree with a stick to fell olives. First, one must not damage the olive itself, as degradation is accelerated if there is bruising or breaking. The goal here is not

Figure 4.12 Images of different types of groves: (left) traditional low density groves in Lesvos, Greece; (middle) high density grove near Canberra, Australia; and (right) super high density grove near Belianes, Spain.

Figure 4.13   Hand-picking and use of small hand-held plastic rakes are common practices throughout the world. Top image, from Sirince, Turkey; bottom two images from Antonij Rupert Estate in South Africa.

to swat the olive off, but to shake the branch enough so that the olive and its neighbors fall off the tree. It is also important when using a stick not to disrupt the fragile bud stalks that, if left undisturbed, will be the site for the growth of the new olives or leaves in the next year. Amateur harvesting can result in a poor harvest in the next season. An expert harvester will place the stick strategically then flick his or her wrist several times to pull the branch quickly back and forth. The best sticks we saw were made of wood (chestnut wood is common in Turkey) and pliable so as to amplify the oscillations and cause the tree to yield up its fruit.

Certain small boutique groves whose focus is on quality and not quantity will use hand-picking as a feature in their advertising.

Hand-picked olives can be collected into wooden or straw baskets, plastic or nylon nets, or into aprons worn by each laborer. Always, the goal is not to bruise the fruit. Knitted nylon nets are said to absorb the energy of the fall of the olive, thus minimizing bruising.[4] In one very small grove we visited in Franschoek, South Africa, that produced exquisite extra virgin olive oil, the grove owner was concerned that olives hand-picked from tall branches where the laborer was up on a ladder might be bruised if they were dropped eight feet to the net waiting below. Her pickers are trained to pick olives into an apron then, when the apron is full, to climb down the ladder and empty her olives into a waiting net. One can imagine how time consuming and therefore costly such a process is. Yet the taste of the resulting oil was testimony to the results of her careful handling of the olives.

Cameo 4.1 Bryan Beverley, Grove Manager, L'Ormarins, Franschhoek, South Africa

L'Ormarins, the largest farm associated with the Anthonij Rupert Winery and Olive Grove, is in beautiful Franschhoek, South Africa. The olive grove was planted in 2001 by Anthonij Rupert just before his death and has become a sentimental link to the family. The cultivars he selected, Coratina, Frantoio, Leccino, and Mission, were planted in a high density format on the steep

(*continued*)

Cameo 4.1   *(continued)*

hillsides of the Franschhoeck Valley. When Bryan Beverley took over as chief horticulturalist at L'Ormarins in 2006, the 2000 olive trees were mature enough to produce enough fruit for harvest. The first thing Bryan did was to reshape the trees, pruning them into a vase-like shape so as to open them up to sunlight and air. The grove uses no herbicides and seaweed extracts for fertilizers. Straw and manure from the estates' horse farm is used as mulch to put more nitrogen back in the soil. Bryan has used a variety of natural organic methods of pesticide control over the years – from use of organic Neem oil as a contact spray for the olive beetle to local prairie chickens in the grove to consume insect larvae in the soil. Sadly, the chickens themselves became food for native lynx cats. In 2015, the year we visited, the fruit was ready to harvest on March 17, St Patrick's Day. The harvest team of about 30 workers from the estate began work early in the morning to avoid the mid-day high temperatures. Handpicking into buckets or nets seemed the preferred method for harvesting on the steep slopes. The four teams processed row by row, faces covered with a chamomile rub to prevent burn from the sun and wind. Bryan encouraged them with promises of an authentic South African Braai if they could meet the picking goal of about 1000 kilos of olives per day or approximately 6000 kilos for the season. We seemed to be the only ones set on edge by the cautionary words to be on the look out for Cape cobras or puff aAdders in the dry grass. While Bryan has considered using hand-held rakes and rotating wands, he fears that they pop the olive and bruise the flesh, and so he continues with the tried and true hand-picking method. The olives are pressed at nearby La Bourgogne Press, are blended into two brands, and a small production of 2000 bottles is produced each year.

4.3.2   Mechanically Assisted Hand Harvesting

It is very common to use hand-held power tools to assist in the harvest of the olives. The most widely used hand-held power tool we saw was a mechanical rake. Powered by gasoline, a rechargeable lightweight lithium battery pack, or simple car battery with a connecting cable, this device consisted of rotating wands on a long pole. The wands or fingers were made of either metal, hard plastic, or soft flexible plastic and could rotate or scissor with varying frequencies. The number and length of wands, the scissoring or raking motion, and the material varies depending on the manufacturer. Examples from a supply store on Lesvos are shown in Figure 4.14.

The limb shaker is a mechanical hand-held machine used to grasp onto medium sized limbs of the olive tree and shake them to release the olives, as shown in Figure 4.15. As with other equipment, it is important that the operator of the hand-held tree shaker be trained as to how to not damage the bark or the buds for next year.

Figure 4.14 Tools used to assist with hand harvesting: two lengths of wooden sticks on the left and an assortment of mechanical rakes on the right for sale, to be used in harvesting olives in Lesvos, Greece.

Figure 4.15 Harvesting olives in Assos, Turkey, using (left) a mechanical rake, (middle) a limb shaker in the hands of agriculturalist Cahit Tunç and (right) a stick.

4.3.3 Mechanical Harvesting

Harvesting olives by hand is extremely labor intensive and expensive – about 50% of the cost of producing olive oil comes from the labor cost of picking.[12] Much innovation in the field of olive agriculture has been focused on developing viable mechanical harvesters that can cut down these labor costs. Where hand-picking may require dozens of workers for every hectare of land, with modern mechanical harvesters, one can reduce that to one to two employees for every 100 hectares. Two broad categories of mechanical harvesters are currently in use. The first category includes machines that progress in a step-wise fashion from tree to tree in a discontinuous manner. The second category includes machines that progress at a constant speed through a grove, harvesting trees grown in dense or super dense rows in a continuous fashion.

Discontinuous harvesters typically work by mechanically shaking each tree trunk one tree at a time. Often, the harvester will travel in the rows between the trees and harvest two trees that are on opposite sides of the row – a so called side by side shaker. The olives are collected into an umbrella or collecting skirt around the base of the tree and transferred into crates or bags as the collector gets full or in moving from one tree to the next. We have seen many different kinds of trunk shakers – each one looking like an unusual insect or strange predator (Figure 4.16). These devices are in use throughout the world and can be used with older and larger trees up to a point. When the tree begins shaking the harvester, rather than the reverse, then it is time to switch to limb shakers or other means of harvest.

Figure 4.16   Discontinuous harvester known as an umbrella shaker in use in Hawkes Bay New Zealand.

Many of the newly planted olive groves in Australia, California, Argentina, parts of Spain, and other parts of the world are enormous, with 2500–3500 hectares per grove. Harvesting such enormous plots would not be possible without mechanized continuous harvesters. Continuous harvesters are used more often in areas where groves have been newly established in a high density (200–600 trees per hectare) or super high density format (often up to 2300 trees per hectare). In these groves, the planting has been carried out with a particular type of harvester in mind, so the distances between rows are optimized to that harvester. Because so often both grapes and olives are grown together, one original grove design was to trellis the olive trees, and then to use grape harvesters that were retrofitted with the olive picking heads. From this early experiment, purpose-built harvesters based on the classic grape harvester design have been developed (one such is shown in Figure 4.17). These harvesters are capable of dealing with substantial slope in the fields, but can't cope with terraced or extremely steep terrain. Most of these harvesters are also designed to travel on roads between groves, if the groves are not contiguous. Thus,

Figure 4.17   Olive harvester based on a grape harvester design, viewed at California Olive Ranch.

a growing and harvesting system is developed, based on planting trees in super high density, pruning them so that they are
of even and relatively low height, and harvesting them with a
machine designed to envelope that hedge.

For high density plantings, with more traditional height trees,
even larger harvesters were developed. One such is the Colossus
harvester, shown in Figure 4.18.[13] Most closely resembling a
moving car wash, the Colossus stands almost six meters tall and
weighs 28 tons. It straddles even full grown trees with large cano-
pies and has specialized olive picking heads, augers to move the
fruit through the machine, high volume bucket containers to collect and deliver the olives to a trailing combine, four wheel drive
and four wheel steering, and moves at a blistering 0.8 km h^{-1}. It
can harvest an average of about 100 olive trees per hour. With a
yield of more than 40 kg olive per tree the cost is approximately
A\$0.06 per kilogram olives, or A\$0.25 per liter of oil.[14] As shown
in Figure 4.19, this is much more efficient than discontinuous
methods.

The advantage is, of course, that many more olives can be
harvested per hour or per day than with hand-picking. Also,
as olives tend to ripen at the same time, fruit is not kept
waiting on the tree until the farmer can reach it for harvest.

Figure 4.18   Two views of the Colossus Harvester at the Greenleaf olive grove
in South Africa.

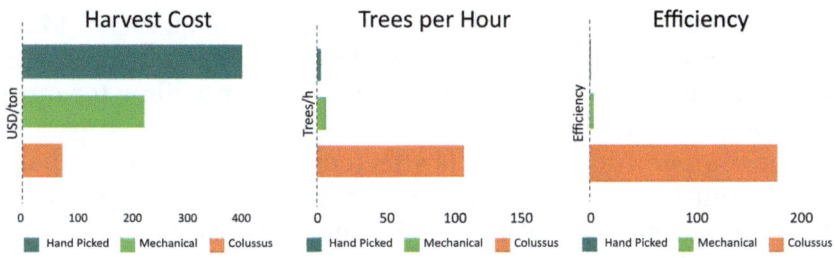

Figure 4.19   Comparison of different olive harvesting methods.

Mechanical harvesters in the large groves need to work 24/7 during harvest times. Most facilities that use mechanical harvesting have their own dedicated presses, and communication between the press and the harvesters can also be coordinated so that if a malfunction happens at the press that might halt processing, with one phone call, the harvesting can be stopped until a part is replaced or a problem solved. This avoids picking olives that might have to wait for hours or days until they can be pressed.

Studies that look at the shaking frequency, rod design, and composition are important for maximizing harvest yield while maintaining quality. One limitation of mechanical harvesting is that these large harvesters can only be used when the land is relatively flat and when trees have been kept below a certain height and within a certain width.

4.4   WHAT CAN GO WRONG DURING HARVEST?

4.4.1   Mistreatment of Olives

Until we invent a machine to teleport the oil out of olives (which according to industry sources and our own knowledge of physics is extremely unlikely) there will remain a simple necessity to gather olives together and transport them to an olive press. The best conditions for transporting the olives include minimizing the pressure on those olives at the bottom of the container (especially for riper, softer olives), keeping the temperature as low as possible, and allowing some air flow around the olives. In many traditional groves, olives are gathered into sacks made of rough cloth or reinforced plastic, and held for several days, which clearly causes significant reduction of quality.[15] Many groves, even in traditional areas are introducing plastic bins which reduce pressure on the bottom layer and allow for more air flow.

The temperature of olives is affected by the external temperature when harvested, of course, but also the conditions while they are waiting. Dark olives in the sun will absorb a lot of light energy and warm significantly. In addition, the olives in the middle of a box or bag can begin to ferment or compost. This can lead to a temperature rise in the middle of the container and increased fermentation. Measuring the temperature of the batch of olives is common in some areas as a way to detect this fermentation.

Of course, any defect in the olive that breaks the skin (crushing, cutting, or heavy bruising) will allow oxygen to contact the interior juices and begin the production of rancidity. Problems become more severe the longer the olives stay in this state and the higher the temperature.

4.4.2 Infections of the Olives

Olives generally have a flourishing microbiome on their skins, just as we do. When held in close confinement, some of these bacteria or fungi can grow out of control, leading to breakdown of fatty acids and other components of the olive to produce flavor defects in the final oil. We will discuss these flavor defects more thoroughly in Chapter 7 on sensory analysis, but would like to highlight some of the defects caused by problems at harvest.

One of the more common defects found in more traditional areas is "fustiness," which is produced by storing olives in deep piles for longer than a day or two.[16,17] Caused by *Pseudomonas* or *Clostridium* bacteria, it gives an unpleasant barnyard or baby diaper aroma to the oil. Under similar conditions, the *Acetobacter* bacteria can flourish and yield a "winey-vinegary" aroma.[18] If the olives are a bit damp, or are slightly crushed, *Aspergillis* or *Penicillium* molds can grow. These fungi break fatty acids not into the pleasant six-carbon species, but into seven to nine carbon compounds that smell like mushrooms or mold. This produces a defective aroma in the oil.[19]

4.4.3 Impact on the Oil

Recalling that olive oil is really a fruit juice, it is perhaps not surprising that problems at harvest lead to defects in the oil. There is no cure for these defects that allows the resulting oil to be extra virgin. Therefore, good harvest practice is absolutely required for high-quality oil.

4.5 HARVEST CELEBRATIONS

Many communities finish the harvest season with a celebration that includes stories, song, dances, seminars, and always wonderful, wonderful food. The Festa de l'Oli is one such festival held in Belianes, Catalonia, in early December. The center of activities is the beautifully restored Maurici Massot olive oil mill – shown in Figure 5.3 of Chapter 5. Behind the mill, with a sweeping vista of the vineyards and olive groves that stretch out and away towards the distant Pyrenees, one can enjoy a typical olive harvester's breakfast of sardines fried in olive oil served on crusty bread with a swig of wine from a community jug (Figure 4.20). Local artisans join school children in marketing crafts and

Figure 4.20   Two of the authors eating a breakfast of whole sardines fried in olive oil at the Festa de l'Oli to celebrate the end of harvest in Belianes, Catalonia.

foods. Throughout the town, homes are decorated with the historic rakes, baskets, and presses and carefully copied out verses from olive folksongs.

Many Turks, both young and old, told us of a traditional olive harvest song – *Zeytinyağlı Yiyemem Aman*. It is easy to find renditions of this song online.[20] What is striking about many public performances of this song is how quickly the entire audience will join in the singing. This is surely an important piece of their cultural heritage. The English translation of the lyrics may be confusing, but this is no fault of the translation. The song, which dates to the middle of the twentieth century, does nothing to celebrate the olive harvest. In fact, it seems to do exactly the opposite.

> *Zeytinyağlı yiyemem aman, (I cannot eat food that contains olive oil)*
> *Basma da fistan giyemem aman. (I cannot wear chintz skirts)*
> *Senin gibi cahile, (An ignoramus like you)*
> *Ben efendim diyemem aman. (I cannot call sir)*

We heard from several Turks that this song was introduced by the Turkish government in the mid to late '40's to convince Turks to shun "old-fashioned" olive oil in favor of modern corn oil. Where did all this corn oil come from? As part of the Marshall plan to rebuild lands ravaged in WWII, foreign aid was given to Turkey. But the aid came with strings attached; in exchange for cash from the US, the Turks agreed to purchase low priced US produced corn oil from its stockpiles of government subsidized corn. Conspiracy theory, or truth? We haven't been able to verify this story, but it certainly adds color to the tradition of struggles to market fats and oils, a substance so fundamental to diet that its purchase is built into a families' habit more than its conscious consideration.

4.6   MOVING ON TO PROCESSING

Once in possession of picked olives, the race is on. Fresh olives are like any other fruit and will spoil fairly quickly if held. This is especially true if they are slightly damaged, as may happen with some harvest practices. Damaged or undamaged, the fruit will certainly last for a few hours without spoiling appreciably. However, the longer they are held, the more risk of spoiling the fruit, and thus the final oil. In the next chapter, we'll discover the amazing process that turns olives into oil.

REFERENCES

1. A. Ferrante, D. A. Hunter and M. S. Reid, *Postharvest Biol. Technol.*, 2004, **31**, 111.
2. A. Dag, Z. Kerem, N. Yogev, I. Zipori, S. Lavee and E. Ben-David, *Sci. Hortic.*, 2011, **127**, 358.
3. I. M. Desouky, L. F. Haggag, M. M. M. Abd El-Migeed and E. S. El-Hady, *World J. Agric. Sci.*, 2009, **5**, 760.
4. *Olive Oil Production*, http://www.mediterraneangardensociety.org/olives.html#2, accessed July 2016.
5. A. Yorulmaz, H. Erinc and A. Tekin, *J. Am. Oil Chem. Soc.*, 2013, **90**, 647.
6. E. Kayesh, L. Shangguan, N. K. Korir, X. Sun, N. Bilkish, Y. Zhang, J. Han, C. Song, Z.-M. Cheng and J. Fang, *Acta Physiol. Plant.*, 2013, **35**, 2879.
7. S. Armenta, S. Garrigues and M. De la Guardia, *Anal. Chim. Acta*, 2007, **596**, 330.
8. L. Salguero-Chaparro, V. Baeten, J. A. Fernández-Pierna and F. Peña-Rodríguez, *Food Chem.*, 2013, **139**, 1121.
9. H. Kocabiyik, R. Lu, M. Seker, I. Kavdir and M. B. Buyukcan, *International Symposium on Application of Precision Agriculture for Fruits and Vegetables 824*, 2008, p. 373.
10. R. Mailer, J. Ayton and K. Graham, *J. Am. Oil Chem. Soc.*, 2010, **87**, 877.
11. E. M. Yahia, *Postharvest biology and technology of tropical and subtropical fruits*, Woodhead Pub, Oxford, Philadelphia, 2011.
12. L. Ferguson, U. A. Rosa, S. Castro-Garcia, S. M. Lee, J. X. Guinard, J. Burns, W. H. Krueger, N. V. O'connell and K. Glozer, *Adv. Hortic. Sci.*, 2010, **24**, 53.
13. *Colossus XL Harvester*, https://www.youtube.com/watch?v=lx-S7s3BzIUk, accessed July 2016.
14. *How Continuous Mechanical Harvesting Transformed the Olive Industry*, http://www.crec.ifas.ufl.edu/harvest/pdfs/presentations/ISMH&HS_Ravetti.pdf, accessed March 2015.

15. O. Koprivnjak, L. Conte and N. Totis, *Food Technol. Biotechnol.*, 2002, **40**, 129.
16. F. Angerosa and C. Campestre, *Handbook of olive oil: Analysis and properties*, Aspen publications, Inc, Gaithersburg, MD, 2000.
17. M. T. Morales, G. Luna and R. Aparicio, *Food Chem.*, 2005, **91**, 293.
18. E. Monteleone, G. Caporale, A. Carlucci and E. Pagliarini, *J. Sci. Food Agric.*, 1998, **77**, 31.
19. F. Angerosa, B. Lanza, N. d'Alessandro, V. Marsilio and S. Cumitini, *Acta Hortic.*, 1999, **474**, 695.
20. *Zeytinyagli yiyemen*, http://www.vidivodo.com/video/candan-ercetin-zeytinyagli-yiyemem-aman/346755, accessed March 2015.

Processing: The Most Important Hour

After harvest, the responsibility for a high-quality olive oil shifts from the olive producer to the manager of the processing plant (if they are not the same person). The producer has been concentrating on harvesting the highest quality olives. Once these olives are harvested, a farmer has very little time to get them to a processing facility if an excellent oil is desired. It's helpful to remember that the olive is a fruit. Like any fruit, if it is left to stand in the sun, it will go bad. One of the most important contributors to a good-quality olive oil is the speed with which the olive gets from the tree to the processing line.

There is an element of collaboration between the olive producers and the processors. To avoid long waiting times for oil production, the press must have sufficient capacity (usually expressed in the mass of olives that can be processed in one hour), and the harvest must be coordinated with other batches of olives arriving at the press. Careful matching eliminates most of the wait, without leaving the process machinery empty. Empty presses are both inefficient and can reduce the quality of the next batch of oil if the processor is not cleaned and the residue in the machinery spoils.

The Chemical Story of Olive Oil: From Grove to Table
By Richard Blatchly, Zeynep Delen Nircan and Patricia O'Hara
© Richard Blatchly, Zeynep Delen Nircan and Patricia O'Hara, 2017
Published by the Royal Society of Chemistry, www.rsc.org

One way of minimizing the time between harvest and processing would be to bring the machinery to process the olives to the grove. Mobile processing facilities/laboratories built inside a truck that can be transported right into the olive grove eliminate waiting times between harvest and processing. A collaboration between Ankara University professor Mücahit Taha Özkaya and Nar Gourmet has successfully used a mobile press with support from local growers in Turkey. In addition to pressing the olives, the facility is set up to analyze the genetic profiles of little known varieties, contributing to the preservation of biodiversity. Similar mobile presses also exist in California and Texas. As this solution can only work for small or medium size groves, we will now turn to the more traditional method of transporting olives to processing facilities.

When the olives arrive at the processing facility, they should also be inspected for disease and pests. The processing of poor-quality olives can contaminate subsequent high-quality olives, so it is usually prudent to separate olives of low-quality for later processing or processing on another line. The olives are separated from leaves and twigs by manual and/or mechanical means such as blowing the leaves and twigs away from the fruit. Washing the olives is a common next step, though some facilities do this only when necessary as their fruit usually arrives very clean and any added water can complicate the next steps. In some climates, when the olives arrive quite cold, the water can be used to warm them to approximately 25 °C or so. Once the olives are cleaned, inspected, and washed, they are ready for processing.

In one facility in Crete, this washing step allowed for a second sorting. Pests such as the olive fly changed the density of their Koroneiki olives enough that they could be floated on water, effectively separating them from the good olives picked early that will sink. Thus, the bad olives could be removed simply during the washing. We didn't see any other producer using this trick, so it may not be general, but a processor in South Africa said olives suitable for becoming table olives floated and in her facility, they were collected and sent off for curing and packaging.

5.1 THE GOALS OF PROCESSING

The processes that occur inside an olive mill are both simple in concept and complex in design. On the simple side: the olives are crushed, the oil is separated from the remainder of the olive fruit, and remnants of fruit and water are cleaned from the oil. Modern facilities can have shiny, loud machines that do this on a large scale with high efficiency, and we will discuss each step later in this chapter, but the truth is that olive oil can be made in a home kitchen. In Şirince, Turkey, we were delighted to be offered a taste of such homemade oil at a small restaurant. Members of the O'Hara-Blatchly family, shown in Figure 5.1, agreed that the oil was sweet, golden yellow, and delicious.

The general method for home production is simple. A mortar and pestle, or another type of grinder can crush the olives, the pulp can be loaded into bags made of cloth or other fiber, and pressed. The bags holds back the solids, while a mixture of olive oil and fruit water can be drained out. Once the water/oil mixture is allowed to settle, it separates into an oil layer that floats on top and a water layer on the bottom. The top oil layer can be scooped out or drained off the top of the two layers. Because the water layer is usually quite black from water-soluble pigments, the mixture looks like a very unappetizing salad dressing that has separated. The water layer has to be discarded, and the solids in the bag can be disposed of, perhaps by feeding to pigs or burning for energy or heat.

This style of processing has been practiced for millennia. Along the way, human ingenuity has been harnessed to make the process more productive, so that barrels or truckloads of

Figure 5.1   Tasting delicious homemade olive oil in Şirince, Turkey.

olives can be ground and pressed. Clever systems for separating the oil layer and the water layer have been discovered in archeological sites dated to 600 BCE. Naturally, the industrial revolution changed the process again, giving first steam-powered, then ultimately electric machines to accelerate the processing, allowing more olives to be processed. Later, centrifuges capable of separating the fruit pulp, water, and oil in two or three phases replaced the presses. It is ironic that modern day olive "presses" no longer press the fruit in any way.

As it turns out, there is more to consider than the simple physical process. Processing is where the flavor of the olive oil is controlled and developed. In this chapter, we will first explore the history of the olive press. Then we can walk through a modern olive oil facility and highlight the decisions that operators in the plant must make – decisions that will affect the quality of the oil produced. These decisions are often about getting the chemistry right, whether the operators know this explicitly or not. We'll explain why the best practices work as well as they do. The overall process critically affects all of the components of a good olive oil: the triacylglycerides that make the bulk of the oil, the aroma compounds that give its smell, the phenolics that give the bitterness and health effects, and the pigments that give it color. Even the final steps in processing will affect the stability of the oil.

It's not surprising that a number of processing plant operators feel a great deal of pressure to do their job well. After an olive producer has spent the better part of a year growing the olives, the oil is produced within a couple of hours. The success of the whole operation depends on the skill of the processor to produce the best yield of the best oil possible. By and large, the plant operators we met were passionate, intensely focused, and felt a great responsibility to the producer.

5.2 THE PHYSICAL STORY OF PROCESSING OLIVE OIL: THE OLIVE PRESS

The steps of olive oil production are the same regardless of the method used: grind up the olive to break up the plant cell walls and burst open microscopic oil storage compartments called vacuoles within the plant cells, stir the ground olive for enough time to develop the flavor of the oil, and separate the oil from the remainder of the material, which is made of solids and a soupy vegetable water. These sound like simple tasks, and indeed if you are doing this on a small scale and don't worry too much about maximizing yield and quality, it's not terribly difficult. Once you start processing a ton or two (or ten or a hundred) of olives, and want to get the best quality oil, control of the process becomes more of a challenge. Not surprisingly, there is an ancient story of industrial design that continues to develop today. For contrast, we'll describe an ancient style press, then move to traditional and modern presses.

5.2.1 Olive Presses Through History

The olive press at Klazomenai, in Urla, Turkey, is a wonderful demonstration of ancient industrial production from about 600 BCE. Introduced in Chapter 1, it will get an expanded description here. The reconstructed olive press is housed in a modest thatch-roofed building, near the site of an iron foundry of the same age. Figure 5.2 shows the reconstruction in action.

Just as with modern processing, grinding the olives is the first step. At Klazomenai, the technique used to grind the olives involved continuously rolling extremely heavy stone wheels or millstones over the olives until the fruit is fully macerated. The fruit was held within a stone basin carved into the bedrock. Two wheels were lashed onto a 10 m tall vertical shaft that was driven by four people. These workers stood on a second story platform pushing against the horizontal bars that connected the shaft to the wheels and walking endlessly in circles about the shaft.

The olive paste, after grinding, was shoveled into fiber bags that were stacked on a platform near the large channel. Over this platform, a disk was attached to an enormous lever, made of four massive timbers lashed together by iron bands. This lever was lifted to allow the olive sacks to be loaded, then a weight was lowered to press out the water and oil. As though the weight of the lever was not enough, huge stones were hung from the end to add to the pressure. A special lifting device had to be constructed to raise this lever.

Judging from the way similar presses have been run more recently, the yield of olive oil can be manipulated in three ways. First, the mass of the olives themselves will cause olive oil to run from the stack of olive paste. This oil was generally of the highest quality. Then, the pressure can be added as described above. This produces quite good quality oil. Because neither of these pressings were heated, they were referred to as "cold-pressed," even if the day was quite warm. Finally, to get the last

Figure 5.2   Ancient Press in Klazomenai, Turkey, originally built in 600 BCE and newly reconstructed in 2000 AD. Top left, millstones attached to vertical shaft at the reconstructed ancient olive mill. Top middle, horizontal bars used to push against to drive the stone wheels. Top right shows paste being transferred into sacks by Cahit Tunç. Bottom left shows the pressing lever and stones used to squeeze the sacks filled with olive paste. Bottom middle shows the oil–water mixture flowing from the mats and being collected into a storage container. Bottom right, author Zeynep Delen collecting oil she pressed at Klazomenai.

drops possible from the mash, boiling water was poured over. As this is no longer cold, the quality usually suffers. This tradition gives the historic terms "first-pressed" or "cold-pressed." These terms are not easy to translate directly into modern practices, except that the temperature is very carefully controlled in a modern press.

Once the liquids ran out of the sacks, they were directed to the central holding pit carved into the bedrock, where, as detailed in the historic cartoons in Figure 1.8, the oil and water collected and separated by gravity, with the oil floating on top. An adjacent pit was connected to the central pit with a channel about two-thirds of the way down, thus allowing the bottom water layer to flow in to it. As the liquid mixture continued to pour into the central pit, water would flow into the leveling pit where it could be removed, allowing for a continuous flow into the central tank. The oil could be ladled into the adjacent unconnected pit for temporary storage. Later, the oil was placed in large clay containers called amphorae, designed to help the oil stay separate from any remaining water and resist breakage while transported.

There are several major differences between this ancient process and the most modern ones – there seems to be no provision for filtration, and there was no control of atmosphere or temperature. But it is also striking how similar this setup is to so many presses constructed over the next thousand years as the technology spread through the Mediterranean basin. The 17th Century stone mill shown in Figure 5.3 is part of the Maurici Massot museum in Belianes, Catalonia. The rotating wheel on the left, pulled by a donkey, and the long lever arm shown on the right, used to squeeze sacks full of olive paste, were nearly identical to what we saw at Klazomenai in Turkey, built 2300 years earlier, with the exception that livestock rather than humans were used to provide the power to drive the wheel.

Stone mills are still in use, especially in traditional production areas around the Mediterranean, but also by modern producers who believe a better flavor results from using this ancient method. Large stone grinding wheels on a pivoting support roll over a stone surface containing the olives. The height of the grinding stone above the surface can be controlled and is usually just a few millimeters. These wheels slowly crush the olives and begin stirring the paste. This style of crushing works the olives in batches, rather than the continuous flow possible with later technology, and allows the olive paste quite liberal access to the oxidizing atmosphere of the press. The risk of contamination and the loss of aroma and phenolic compounds from the oil have encouraged many operators to modernize. Yet, the stone mills are very durable. Figure 5.4 shows a 100 year

Figure 5.3   17th Century stone mill (left) and press (right) in the Maurici Massot museum in Belianes, Catalonia.

Figure 5.4   100 year old millstones (left) and hydraulic presses (right) currently in use in the Moli Duran olive press in Catalonia.

old stone mill still in use in Belianes. Aside from their pictur-
esque appearance, some modern producers maintain that this
method reduces the intensity of a very bitter oil. The massive
stone wheels made a deep rumbling sound as they turned on
the stone base. Instead of a modern separator, the press operator shoveled the paste into fiber sacks and pressed them using
an old hydraulic press that was identical to those we saw in
the Museum of Industrial Olive Oil Production in the village of
Agia-Paraskevi in Lesvos, Greece.

5.2.2 Modern Processing

Like many industrial processes, olive oil production was transformed by the invention of the steam engine. It is not hard to
imagine how excited producers were to gain the assistance of
steam power to crush the fruit and press the oil out of the paste
filled sacks. In some places, steam was generated by burning the
oil-rich solid waste (including olive pits). A modern mill is run
completely by electricity, leading to much greater control over
the speed and power available for each step. Running a mill is
still a big job, but it can be done by many fewer people.

We should note that these processing systems (still called
"presses," to honor the tradition despite the fact that no actual
"pressing" occurs) come in many different sizes. Small presses,
capable of processing 1 kg per hour, sit on a workbench in a
garage. The largest presses, capable of processing several tonnes
per hour, can be found at large processing centers like Boundary
Bend in Australia or California Olive Ranch in the US. The choice
of size is very important. If the press is too small, then olives will
begin to accumulate during the harvest, leading to poor results.
If the press is too big, the price will be higher than necessary,
and it will be hard to keep the components working continuously, leading to off-flavors as the residue in the machine spoils
between batches.

The process requires grinding the olives, mixing the paste for a
specified time (malaxing), making a first crude separation in the
decanter, and often a second finer separation in the centrifuge.
The finished oil is sometimes filtered. These steps are shown
schematically, starting with the olives, in Figure 5.5.

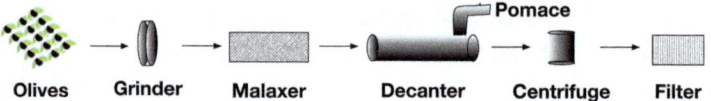

Figure 5.5   Schematic diagram of olive oil processing.

After good-quality olives are selected, the two factors that have the most impact on the quality of the oil are the temperature of the overall process and the speed with which the steps are carried out. The temperature of the process is, not surprisingly, determined mostly by the temperature of the olives as they enter. The grinder will heat the olives, as will the decanter. While the malaxer is usually jacketed with a water sleeve of controlled temperature, the capacity to change temperature in a short time is limited. This is one of several reasons that olives must be kept as cool as possible during harvest. Many processors are routinely taking the temperature of olives as they arrive.

The speed of the process relates mostly to the malaxer, and to some extent the decanter, and will be discussed below.

5.2.2.1   Grinding the Olives.   After washing, grinding or milling the olives is the first big step towards producing oil. It may be a surprise that the whole olives can be ground, but even ancient stone grinding wheels could be adjusted to crush the olive pits along with the fruit to enhance the processing. With modern processors, the fruit is most often crushed using a powerful grinder that runs metal blocks across a strong sieve while the olives are forced into the chamber. This operates a bit like a flour sifter or grater, but with much more force. This machine, called a "hammer mill," very quickly produces a paste from the olives, which can be pumped to the next stage of the process. The speed of this process means that the following process can be controlled more carefully, which is quite important. The press operator chooses the size of the holes in the sieve and the design of the mill (there can be one set of hammers and one screen or two) depending on the type of olives expected. In addition to the speed, hammer mills have the advantage of being completely

Figure 5.6 Looking inside a hammer mill. Left: opened for end of season cleaning at The Olive Press in Sonoma, CA. Right: at the end of a day's work at the BioOleics Mill in Belianes, Catalonia.

enclosed, preventing contamination, and can operate continuously. Figure 5.6 shows the interior of hammer mills when opened for cleaning.

Similar to the hammer mill, in that it is a continuous crusher, is the knife crusher. In contrast to the hammer mill, the knife mill spins sharp blades within a chamber, slicing rather than crushing the fruit. Many knife mills are variable speed, allowing variation in the degree to which the fruit is disrupted.

Stone mills are still in use in traditional areas and have their fans in New World presses as well. Grinding the olives and beginning the mixing process while open to the air will degrade the more bitter antioxidants more than a closed mill such as the hammer mill, so might be appropriate for olives that would otherwise give a very bitter oil. An example of a stone mill is shown in Figure 5.4.

Removing the pits from the olives is an optional step before grinding. Many mill operators say that the sharp crushed pits help in the maceration of the olive fruit necessary to achieve the highest yield of oil. Others say the pits can make the oil too bitter, and there are machines that can remove the pit prior to processing, though relatively few processers do this.

Cameo 5.1   Josep Ramon Morera, co-owner BioOleics, Belianes, Catalonia.

Josep Ramon Morera may be the mayor of the tiny village of Belianes, Catalonia, but his jeans and farm jacket signal his priorities – at heart, he is a farmer, olive producer, and entrepreneur. His company BioOleics produces an award winning organic extra virgin Arbequina olive oil, marketed under the brandname Camins de Verdor. He manages to embrace traditions of olive farming, harvesting from older groves while experimenting with new hybrid cultivars planted. The BioOleics olive press aspires to carbon neutrality – and Josep Ramon was happy to show us reports detailing his carbon footprint. He collaborates with Agromillora, the world's largest nursery and a leader in the development of new cultivars suited to super high density farming practices. The 2014 Umami EVOO, made from early harvested Arbequina olives, was one of our favorites. His excited "Let's Go!" rallied us as he led us from one site to another.

5.2.2.2 Malaxation.   The next step in production is called "malaxation" or beating. The word may seem strange, but it comes from Latin or French (the Latin *malaxo* means to knead or soften, and the French *malaxare* means to mix). It is a slow and thorough process of stirring the olive paste. The process of malaxation may look both boring and pointless. After grinding up the olives, why not just spin them in a centrifuge and collect the oil? However, from the perspective of the yield and quality of the oil, it is the second most important step of the process, after the selection of good olives.

There are two major aspects of this process, one physical and one chemical, occurring simultaneously during malaxation that are crucial to the success of the product. The physical aspect of the process results from the continuous beating of the oil in the malaxer for anywhere from 30 minutes to an hour. Since the grinding of the olives in the hammer mill does not break apart all of the cells in the fruit, the slow stirring tears up any of the remaining cell walls and allows time for the tiny droplets of oil to find each other and make larger droplets. It's important to have larger drops of oil, because when it comes time to separate the oil from the water, they can coalesce out of a pretty thick mush. For a long time, the consolidation of the oil was considered the main benefit of malaxation.

Within the last couple of decades, researchers looking at the development of flavor in olive oil have proven something that astute olive producers have suspected: much of the flavor of the olive oil is actually developed during malaxation. For an oil to be truly high-quality, it must have a noticeable positive aroma (that is, smell fruity, grassy, or the like). Without this, it is not of the highest quality. Natural chemical reactions that occur in the olive paste during malaxation multiply the amounts of aroma compounds dramatically, leading to most of the wonderful smell. Another important process taking place in malaxation is the development of bitter phenolic compounds in the oil (compounds related to oleuropein, which also convey several health benefits). There is an even more complex set of processes associated with these compounds that both enhance the supply of phenolics and degrade them. The chemical story of these transformations will be told in the next section. First, let us finish a discussion of the process itself.

Just as in the historic presses, the malaxing process responds to temperature changes. The oil tends to thin out and flow more easily (become less viscous) at higher temperatures, and a free flowing oil is much easier to handle than a thick viscous slush. Coalescence of the oil micro droplets into a physically separate layer is also accelerated at higher temperatures, letting the same machine process more olives per hour. While efficiency and high yield are still important today, most malaxers have a tempera- ture controlled water jacket to prevent overheating of the paste, either from the friction created by the paddles moving through the paste or from the ambient temperature. Higher temperature has its risks: if the temperature of the paste goes above about 35 °C, degradation reactions take over. High temperatures will also slow down some of the reactions needed for flavor develop- ment because the enzymes needed to catalyze the reactions are destroyed by heat.

Of course, low temperatures can also be a problem. If the tem- perature of the paste goes much below ~25 °C, the reactions for flavor development will slow down, resulting in an oil with little flavor. It also becomes quite difficult to get a good separation of the oil from the water layer, resulting in a lower yield. Since extra virgin olive oil (EVOO) is usually processed at or below 27 °C, it is a bit redundant to include the words "cold-pressed" on a label.

Responsible plant operators must be concerned about both yield and quality, and so the decision of what temperature to use is a rather difficult one. Too high and the flavors can be degraded, too low and the oil doesn't separate from the paste. The optimum temperature also changes with the cultivar and the water content of the olive, which means that the temperature may need to be adjusted several times a day as different olives are processed. In most plants, the temperature of the paste is regularly measured throughout the day and night and the water jacket temperature adjusted if the paste was too hot or too cold.

The control of malaxing time is equally important. As men- tioned above, flavor development is accomplished by balancing productive and destructive reactions that happen simultane- ously. This is like any recipe in which a chef is balancing the need to correctly cook ingredients without burning them. Too short a process leads to incomplete flavor development, and too long leads to lower phenolic content and potentially some off flavors. In addition, with long malaxation times, surfactants can be released from the olive that turn the mixture into a mayon- naise like emulsion, from which the oil will not separate. Each cultivar has its optimum time, which good plant operators get to know with experience. It is worth noting that the mixing time really starts as soon as the olive paste leaves the grinder. It can take 15–30 minutes to fill a malaxer, so the design of the malaxer and how it is run will determine how evenly the fruit is stirred. A manager or operator will oversee this process, visually inspecting the paste during its time in the malaxer to check on the progress of the separation. When pools of oil begin to coalesce on top of the paste, the manager often declares it done and the paste is moved onto the next step.

5.2.2.3 Centrifugal Separation of the Olive Paste in the Decanter. Once the malaxation is finished, the mixture must be separated, ideally with oil ending up in one container and everything else in another. This is a bit harder than it sounds, but this is not far from what a modern separator does. Even more amazingly, it does so continuously – a major improvement from the hydraulic or lever presses that squeezed one batch of olive mash at a time. Modern separators are also referred to as decanters and can process olive mash for hours at a time without pause.

The paste has three major components: a relatively dense solid-like material which has the flesh and pits of the olives in it, a moderately dense layer which is largely water with dissolved material in it, and a less dense layer with the oil. Like a front loading washing machine going through its spin cycle, the decanter spins the mixture horizontally at speeds of about 3000 rpm, causing the more dense components of the mixture to be forced to the outside while the less dense components of the mixture stay closer to the center. With the additional forces acting upon the mixture, the layers separate much more rapidly than in older gravity based decanters. This allows the production of oil to keep up with the pace of grinding and malaxation.

The choice of decanters is roughly divided into what are called "three-phase" and "two-phase" separators. The traditional process can be described as a variant of the three-phase process, as the olive paste was pressed in filter bags, retaining the solid phase (1), and the resulting liquid was separated into the water (2) and oil (3) phases. The more modern centrifuges do the same job with the mixture being fed into the spinning cylinder, the phases separating rapidly, and ingenious augers, nozzles, and gates that efficiently pull off each phase separately. An example is shown in Figure 5.7.

Figure 5.7   Large decanter at California Olive Ranch opened for cleaning. The auger normally fits inside the horizontal cylinder behind it.

This extremely clever machine can be tricky to set up as the spinning speed of the cylinder (usually about 3000–5000 rpm), the speed of the auger (a few rpm faster than the cylinder), and the relative heights of the liquids and their outlets can all be adjusted. Moving the solids through too quickly by using a faster auger helps the throughput but leaves more oil in the solids. Likewise, the location of the exit gate for the oil relative to the oil–water interface can be adjusted to give clearer oil, but the yield will be lower. It can take considerable experience to manage the settings on this machine.

The most modern plants use a two-phase rather than a three-phase separation. A two-phase separator carries out the same separation using the same principles but combines the water and solids into a slurry, which is removed without further separation. It is designed with a single liquid outlet for oil. The water and pomace form a wet slurry, which is separated as one item. Mechanically, this is easier as the slurry flows more easily than the solid pomace, and this type of machine uses less electricity as a result. In addition, because much less water is added to the process, the volume of waste is much smaller, giving an additional cost saving in disposal.

How one deals with the wastewater and solids is a huge problem for the olive industry with no single good solution. A full discussion of these issues can be found in Chapter 10. Some presses will take advantage of the fact that there is some residual oil in the pomace that can be re-extracted to make lower grade pomace oil. Alternatively, the slurry can be piped into trucks and transported to remote sites for repurposing.

5.2.2.4 Centrifugal Polishing of the Oil (If Desired). In most presses, one final centrifuge stands between the olive and the finished oil. The oil as it comes out of the separator is usually cloudy and has some tiny droplets of water suspended in it as well as bits of solids. A vertical centrifuge, which traces its heritage to those used to separate cream from milk, spins the oil at a higher rate of 5000–7000 rpm, separating much of the remaining solids and water. Additional wash water is generally added to aid the separation. This washes away any remaining water-soluble particulates (see Figure 5.8).

While the resulting oil might still not be perfectly clear, it is now free of most of the undesired material, especially the remaining vegetable water. There are, of course, those that like a cloudy oil, right out of the final tap. It is delicious, but because the tiny bits of water and olive flesh still have enzymes capable of degrading the oil, the shelf life is very much reduced. Once you open the bottle, you should count the remaining life of the oil in days or weeks. Bottles of oil right off the press can be purchased from local vendors in olive growing regions. If you ever have the opportunity to enjoy this sensory treat, do so!

Most producers take one or more additional steps to provide a clarified oil with an extended shelf life.

5.2.2.5 Filtration (If Desired). There is quite a difference of opinion regarding the use of filtration in olive processing. We have spoken to processors who swear by it, and those who swear at it. The downsides are several: the yield will decrease as some of the oil will be lost to the filter material, some of the critical taste and health giving substances may be preferentially absorbed by the filter, the oil may be exposed to the atmosphere for an additional short period of time, and the filter materials are expensive. There are additional labor and time spent in keeping the filter running smoothly.

Figure 5.8 Polishing centrifuge at Morgenster Estates, Somerset, South Africa.

Some plant operators believe that the small losses of oil to the filter and the filter expenses will be offset by the extended shelf life of the oil because filtering dehydrates the oil, removing residual water and water-soluble enzymes. Concerns about exposure to air can be mitigated by keeping the filtration system closed to the atmosphere or blanketing the filter with an inert atmosphere. Proponents of filtering assert that the shelf life of the oil is considerably increased. Some processors filter before storing the oil in large tanks and others filter before bottling. Still others do both.

When measured carefully, one finds that the filtration process does decrease the amount of some phenolics in the oil.[1] The reduction is not uniform for all of the compounds and is substantial but not overwhelming (perhaps up to 20%

reduction in some compounds). Whether or not this is acceptable no doubt depends on the phenolic level in the raw oil and the goal of the producer. It is also possible that the phenolics removed were not in the oil, but in the water suspended in the oil.

The material that the filter is made of will play a role in the filtration process. Cellulose filters may absorb more of the phenolics and antioxidants but are less expensive. These filters may have to be changed frequently and are not re-useable. In some circumstances, this is labor intensive as it must be done by hand. Nylon or other polymer filters will bind less of the valuable phenolics and antioxidants. They are often re-useable but are more expensive. If they clog up during the processing, they can be cleaned manually or with a back flush.

Whether or not the oil should be filtered at this point in processing depends upon many factors, most importantly the nature of the olives that are being milled, the degree to which the processing has been effective at separating the oil and the fruit water, and the final taste of the oil that the press operator desires.

We have tasted prize winning filtered Spanish oil that was quite bitter and had a high level of phenolics and prize winning South African non-filtered oil that was not anywhere near as bitter but still had a high level of phenolics. Both were from *arbequina* cultivar. Growing conditions will make a big difference and, in particular, irrigation will especially affect the flavor of the oil. Many of the smaller more mature groves in Spain were not given as much water as the trees in the hot South African climate. In fact, we were told the Catalonian grove would be given water once in the season by flooding the grove and letting the water be absorbed into the clay base. The grove was given no water after that. So, sometimes, filtering is exactly what one should do and other times it is precisely what you should not do. The attentive olive producer will experiment with the oil produced locally and decide whether or not to filter.

5.2.2.6 Final Step in Olive Oil Processing – Racking. Most often, the newly produced oil is transferred to stainless steel holding tanks and covered with a blanket of inert nitrogen or argon gas to prevent oxidation. Small presses may have several 50 L holding tanks while larger facilities can have scores of 1000 L holding tanks. In some facilities, the holding tanks are kept in air-conditioned spaces between 13–18 °C to further reduce degradation. The newly pressed oil is given weeks to months to settle in a process called racking. During this time, residual water that may have settled down to the bottom of the tank is pulled off through valves in the bottom of the holding tanks.

If possible, each cultivar is stored in its own holding tank. In large facilities, oil pressed earlier in the season will be kept separate from oil pressed later in the harvest, even when they are the same cultivar. In the largest groves, the olives of one cultivar grown on one part of the estate may have a different flavor from the same cultivar from another part, and so are kept separate. In these large facilities, the holding rooms for the dozen or hundred giant tanks are called tank farms.

5.2.2.7 Blending. Just like wine, olive oil sold in the stores can be a single cultivar or a blend of cultivars. The personal taste of the consumer will dictate which one is "better," but each cultivar has a distinct flavor profile that can change from year to year and from region to region. Mill operators tell us the flavor profile of the oil develops during this racking period, and if the oil is to be a blend of cultivars, such as a typical Tuscan blend of 35% leccino, 25% coratina, 40% frantoio, the blending is usually done after the flavor evolution has stabilized. Expert tasters, often including the plant owners, taste the racked oil from each tank and decide exactly how much of each cultivar to mix together to produce the desired result. It is most common for the blending to happen just before bottling, but it is also possible to blend and store the oils that way.

The Fairview Olive Mill in South Africa allows customers to sample several cultivars, decide on their own blend, and to mix according to their own tastes. The information is stored in a computer from year to year so the customer can go back to the mill at a later date and be reminded of the composition of the last oil purchased and receive a new product without bothering to go through the trouble of re-tasting and blending the individual cultivars.

Beyond the holding tanks, another world exists where decisions need to be made about storage, bottling, transport, and marketing. These issues again are critical to the quality of the product you consume and are discussed in Chapter 6.

5.2.2.8 Improving Processing. Water in the fruit can be a problem during malaxation. When the water content of the olive is too high, stable emulsions of water and oil may be created that prevent the clean separation of the water and the oil. If the mill operator sees that the olives just delivered have a high water content – for instance, if it has rained just prior to harvest or if the cultivar is known to have a high water content – some talc (magnesium silicate) may be added to bind up the water and help with the separation of the oil. The addition of talc is as old as the hills, and even the most stringent rules for the preparation of the highest quality oils allows its addition for the production of the highest quality oil. The wet talc will be removed with the fruit solids and not affect the flavor of the oil.

The barriers to collecting all of the oil inside the olive include the robust structure of the cell walls that make up the olive. While one would think that grinding the olive to a fine mash in the hammer mill and stirring it for nearly an hour in the malaxer destroys all of the internal structure, believe it or not, some of the cells are left fully or partially intact. Separating out the solids at this point will then sweep up these intact cells with the pomace and that residual oil will go to waste. Some producers will re-extract the pomace to get pomace oil, which is still economically valuable although not as valuable as EVOO.

To maximize their yield of oil, some producers have decided they will help nature along by adding an enzyme boost during the grinding step. The enzymes added are natural enzymes that can help with this process of breaking down cell walls. With the cell walls gone, more oil can be collected and some of the other quality indices improved as well.[2,3] Since the enzymes are water-soluble, they should never show up in the oil.

Others believe that adding anything to the oil besides water and talc renders the oil produced no longer extra virgin. Indeed the IOC definitions seem to preclude addition of anything to enhance the yield of the oil but the use of enzymes is not specifically excluded. Still, we know that in many areas the guidelines evolve to keep up with new understanding of how olives grown in different parts of the world might have slightly different compositions but still be perfectly good and authentic.

5.3   WHAT'S THE INSIDE STORY OF PROCESSING?

If you have never visited a modern olive press during the harvest, the tour can be quite a feast for the senses. The presses often run continuously for almost two months – some of them running night and day. Olives stacked in giant crates or large sacks are turned into a beautiful stream of golden green olive oil all under one roof and within the space of a few hours. The press contains loud, powerful machines, overseen by plant managers and staff and sometimes assisted by central computers. The modern machines are glossy stainless steel and connected by complex metal or plastic conduits. The smell is that of the freshest olive oil.

The story of what goes on inside the macerated fruit is at least as interesting and complex as the story of the machines themselves. Since this is, as we have said, where much of the flavor is developed, our next task is to move inside and try to imagine what it's like to be one of the molecules that will end up in the oil. We'll start by examining the cellular structure of the olive, and then move to a smaller scale to discuss the journeys of the molecules that end up in the oil.

5.3.1   What is the Underlying Cell Structure of the Olive Fruit?

An olive at optimal ripeness is only about 15–25% olive oil. This means that 85–75% of the fruit needs to be separated from the olive. Some of this remainder has other uses, so we shouldn't think of it as waste, but it won't end up in the oil. Olive solids, such as fibers in the flesh, and the olive pit fragments remain solid through the process, while the water in the fruit will mix with water added during processing. Neither of these are good for long-term olive oil storage and are removed. Much of the extra material comes from the plant cells that make up the olives themselves.

The oil is predominantly in the flesh of the olive, which is composed of individual tiny cells, invisible to the eye. The flesh is protected by a tough skin with a waxy outer layer. At the center of the fruit, there is of course the pit, also called the stone or the bone of the olive, whose sole purpose is to finds its way into the soil and grow a new olive tree.

Each plant cell within the olive fruit has a rigid cell wall and an inner cell membrane around the entire cell that holds the insides (called the "cytoplasm"). The cell wall gives the cell strength and shape. The cytoplasm contains the majority of the water-soluble nucleic acids, enzymes, nutrients, and waste products. In addition, these cells have internal structures that further sequester the genetic material contained in the nucleus, or mitochondria that supercharge cells to further break down nutrients to ATP and CO_2. Plant cells also have specialized structures called chloroplasts for localizing the photosynthetic apparatus, including chlorophylls and enzymes. Most relevant for this story, as is also shown in Panels A and B of Figure 5.9, mature olive fruit cells have particularly large interior storages of oil contained in structures called vacuoles or free as lipid droplets. These vacuoles and droplets can take up much of the volume of the cell.

5.3.2 What's Going on During Processing?

The world as experienced by a molecule is a very strange place. Molecules don't have senses in the way humans do. However, they do respond to their environment in ways that are vaguely familiar. Of course, molecules don't have eyes and are too small to "see" but still, they are not blind. Some molecules are light sensitive, absorbing light one photon at a time and changing shape or reacting with nearby molecules as a result. We know molecules cannot hear, yet if the temperature is increased, they will respond by vibrating a bit more vigorously. Vibrations are what our ears pick up as sound, and while the molecules are not really "hearing" the vibrations around them, they can absorb energy from these vibrations. Likewise, molecules don't have a sense of

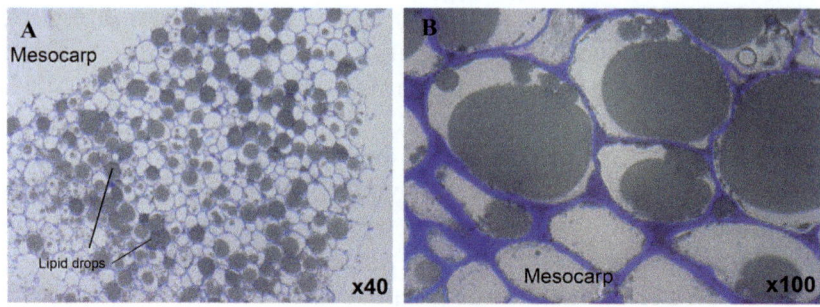

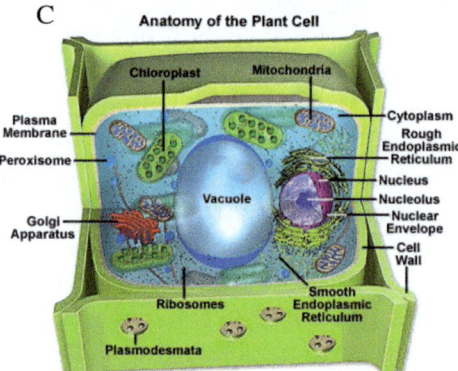

Figure 5.9   Panels A and B show microscope images in which the plant walls
are stained blue and the lipid drops show as dark gray. Reprinted
from Eduardo Lopez-Huertas, Luis A. del Río, Characterization of
antioxidant enzymes and peroxisomes of olive (*Olea europaea* L.)
fruits, *J. Plant Physiol.*, **171**(16), 1463–1471, Copyright 2014, with
permission from Elsevier. Cell structures of the plant. Panel C is
a cartoon of the plant cell. Note the double cell wall that protects
the fragile contents of the cell, the chloroplasts, and the large
vacuole in the center of the image (http://micro.magnet.fsu.edu/
cells/plantcell.html).

touch, but they do respond to the shapes and polarities (charges)
of other molecules around them and hold more or less tightly to
those others depending on how complementary they are.

   With a little bit of imagination, we might be able to follow the
journey of some characteristic molecules as they experience the
chaos of olive oil production. Because molecules of different
types have different structures, they will experience this journey
in different ways. Because they all follow the rules of chemistry,
we can use those rules to explain why the production process is
so important.

The main classes of chemicals in olive oil are what make it unique. We'll describe these briefly below and give representative samples of each class so that you will know what they look like.

5.3.2.1 Triacylglycerides. Triacylglycerides (TAGs), also known as triglycerides are rather large molecules, made from three fatty acids and one glycerol, as described in previous chapters. They are by far the largest proportion of the oil and give the oil its caloric content. A picture of the typical TAG, triolein, shown in Figure 5.10, is remarkable for its large size and scarcity of red and blue patches that would make it polar. Instead, the predominant feature in this molecule is the wide expanse of green that signals this molecule will be very nonpolar.

When the olives are brought to the mill, the TAGs are all stored in microscopic vacuoles in the cell. From the perspective of the olive, the purpose of these vacuoles of calorie-dense TAGs is to feed the olive pit during its transition from seed to shoot. For this reason, the TAGs are kept separate from the rest of the olive cell. While the olive is intact, the vacuoles stay intact, and we cannot easily collect the oil. The first sense that TAGs might get (if they had senses) that something is up comes when the olive is ground up in the mill and the paste unceremoniously dumped into the malaxation vat with powerful stirring devices.

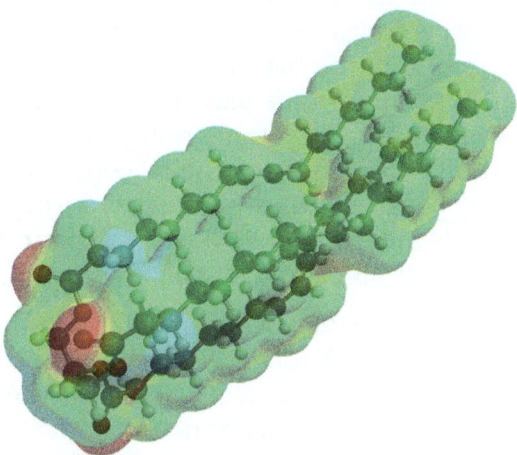

Figure 5.10   Structure of triolein.

As previously mentioned, during malaxation, most of the cells break open, as do the membranes surrounding the vacuoles. Once liberated from the vacuoles, suddenly the TAGs find themselves in contact with the rest of the olive, which is mostly water. The natural tendency of the TAGs will be to find other TAG molecules and coalesce, making separation easier. As the membrane of the vacuole is disrupted, the TAGs spill out and come into contact with other molecules in the cell. One possibility is that the contents of one vacuole meet the contents of another. It's not hard to imagine that they just join two droplets into one, making a bigger droplet. As this process continues, we begin to get droplets that are big enough to see without a microscope.

This is good news because it takes quite large droplets to separate the oil from the rest of the olive materials. This process will continue until there is an easily visible continuous layer of oil. Some operators use the appearance of this oil layer on top of the paste as a sign that the malaxation is finished. There remains lots of murky water phase (solids and liquid together), and when malaxation is done, the water phase needs to go. That's the signal to move to the next step. Before we do, however, let's talk about the development of flavor in the oil. Since TAGs have no taste, the flavors must come from other molecules.

5.3.3   Developing the Flavor During Processing

Olive oil flavors come from a variety of sensations: aroma that is sensed in our nose, bitterness sensed primarily on our tongues, and pungency sensed in the back of the throat. In addition, the color of the oil is an important part of its enjoyment. The raw olives, and thus the olive mash that enters the malaxation, have a flavor profile quite different from the oil collected at the end of the process. Much of the change is effected during malaxation. Aroma compounds are developed, bitter compounds are modified, and pigments are altered, all during malaxation.

5.3.3.1   Developing Aroma Compounds.   This journey to produce aromas starts with a TAG that contains at least one polyunsaturated fatty acid (PUFA). This type of TAG forms the minority of the oil present, but we don't need much material to make a big impact on the aroma. Even producing a very fragrant oil leaves the TAG almost completely intact.

To make any aroma molecules, we need another actor: some special natural enzymes. In the growing olive, these flavor enzymes are mostly held separate from the stored TAG molecules. However, when the cells are crushed, the enzymes can mix freely with the TAG. Enzymes act as catalysts, which mean they can speed up chemical reactions that might otherwise take too long or require harsh conditions to proceed. The aroma producing enzymes, which are large proteins, will envelop the TAG molecules in the olive paste and begin to break apart some of the bonds that hold them together, making aroma molecules. These enzymes are very water-soluble, so are not found in the oil.

Of the many enzymes in the olive cells, the lipoxygenases are arguably the most interesting of the enzymes here since they break down fats. The name is derived from three simple parts: "lip-," which stands in for lipid and means the enzyme binds to nonpolar compounds; "-oxygen-," which means that the enzyme also binds oxygen and uses it in the chemical breakdown reaction; and "-ase," which in chemical naming conventions means that this is an enzyme or catalyst. There are two important lipoxygenases in the olive flesh, whose job in the intact olive is to make the olive smell like an olive (important for a plant that wants animals to eat the olive and distribute the seeds). Any TAG containing a PUFA that also encounters this enzyme along with a molecule of oxygen will find the last five or six carbons of its chain rudely, but efficiently, chopped off. It is these five to six carbon long fragments that are turned into molecules that smell like olive fruit, green grass, or green apples, or other similar delicious things. Despite the efficiency of the enzymes, the process of developing smells can take quite a while to get just right, so the malaxation is usually run for the better part of an hour. The longer they are in contact with the TAGs (at the right temperature), the stronger the aroma developed – usually much stronger than is present when the olives arrive.

As a large molecule with a specific, almost mechanical, function, enzymes may be a bit more sensitive than other molecules. To carry out their jobs efficiently, they must have a very specific structure that is flexible enough to let in the molecules that need to react, but rigid enough to force the reacting molecules into the correct shape. Since their role as catalyst means that they must return to their original shape at the end of the reaction, any shape change is dangerous. If the temperature of the surroundings becomes too high, the enzymes begin to unfold or denature, and a denatured protein will stop working. Being outside of the cellular environment is hard enough on an enzyme and some don't even survive that transition. Heat only makes things worse. Practically, temperatures much above about 30 °C are too high for the enzymes to work well, which is why cold-processed olive oil tastes so much better than oil processed at a higher temperature. Likewise, at too low a temperature, the aroma development may be too slow, as are other required physical changes.

Fortunately, we don't need to see a lot of this reaction. In contrast to the TAGs, aroma compounds are present in relatively small quantities in the oil and are physically much smaller (about one tenth the size). The smaller molecules are much more likely to go from liquid to vapor, allowing them to travel to our nose and excite the receptors there, giving us that wonderful sensation (if it's a good oil), or that sense that something is off (if it's a bad one). Small in quantity, but big in impact, aroma compounds help to define the quality of the oil. Figure 5.11 shows the chemical structure of a compound that gives a grassy aroma note to olive oil. Note that this molecule is very nonpolar (it is mostly green), so dissolves well in the oil.

Olive oil contains hundreds of different compounds in its aroma profile, and every batch is very slightly different. There are some similarities based on cultivar and ripeness, but even the same cultivar may have a different mixture, and therefore a different aroma if grown in a different climate.

Figure 5.11   Structure of 2-hexenal.

5.3.3.2 Issues with Aroma Development. Because the aroma develops slowly during the malaxation, it is important to continue the process until that development has matured. The process does require oxygen. However, the amount of aroma compound is very small, so the quantity of oxygen is so low that the paste actually contains enough to complete the aroma development.

It is easy to lose aroma molecules by evaporation, especially if the temperature of the process is too high. Evaporation rates do increase with temperature, which is one good reason to keep the temperature moderate. Many modern malaxers are also sealed, keeping in the aromas.

5.3.3.3 Phenolic Compounds. Phenolic compounds are often referred to as polyphenolics or antioxidants. Some typical phenolics are shown in Figure 5.12. Many of them are also bitter to various degrees (and one of them gives the peppery finish known as "pungent"), as described in Chapter 7. Some of these compounds have other health benefits, as we will discuss in Chapter 8. These molecules have intermediate polarity, and are quite soluble in oil but also quite soluble in water, posing an interesting challenge for the olive producer. Most olive phenolics do not make it into the oil – one careful study showed that less than 1% of the phenolics in the olive made it into the oil.[4]

The development of the bitter flavor begins with a pair of compounds called oleuropein and ligstroside. Despite their very different names, they are very similar compounds, with only an oxygen atom at one site to tell them apart. Each one tastes extremely bitter and accounts for the taste of olives before curing (see Figure 5.13). Because oleuropein is more common, we will follow it into the processing. As discussed later, given a choice between water and oil, oleuropein stays resolutely in the water phase, due to its large number of polar groups.

Oleuropein is abundant in green olives and in olive leaves. When the olives are crushed, the oleuropein is broken down, as shown in Figure 5.12. The breakdown begins with a simple reaction, accelerated by an enzyme called β-glucosidase, in the flesh of the olive that can remove a sugar component from the oleuropein to make a new compound that is slightly less bitter, but which is much more soluble in oil. This compound, called by the

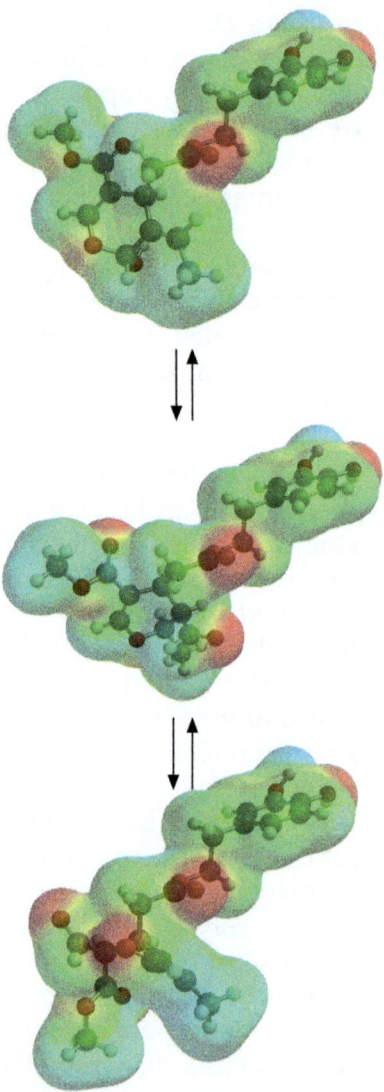

Figure 5.12   Chemical breakdown products of oleuropein. From the top, oleuropein aglycone, closed form of DHPEA-EDA, and an open chain form of DHPEA-EDA. These are only a few of the possible compounds produced.

ungainly name oleuropein aglycone, is still very bitter, has many of the health effects of oleuropein, and dissolves well in oil. It may seem strange that a very bitter compound with an attached sugar becomes less bitter when the sugar is removed, but this is quite common.

Figure 5.13 In the spirit of research, authors Blatchly and O'Hara have tasted uncured olives and do not recommend the experience as pleasant.

Once you remove the sugar from these compounds, many different kinds of chemical rearrangements are possible. Each of these compounds still contains hydroxytyrosol (that conveys the antioxidant properties) but, in each of the three structures, the rest of the molecule has been rearranged to form a different structure. The same kinds of reactions can occur with ligstroside. Figure 5.12 shows three of these compounds derived from oleuropein, which are meant to represent a very large family of related structures. If you include the different shapes attainable for each of the carbons in the ring, the total set of compounds is dizzyingly large: about six important compounds and dozens of minor ones.

If the production of these bitter compounds from oleuropein increases their presence in the paste, then longer malaxation times should raise the levels of phenolics. There is another side to the story, however: the reaction of phenolics with oxygen. When the hydroxytyrosol portion encounters an oxygen molecule, it reacts to produce an unstable modified form. This is the normal response of antioxidants, and it is important to remember that the compound is destroyed as a result. This unstable product of reaction with oxygen will continue to degrade, producing forms that make them no longer soluble in the oil, so they are also carried away with the pomace.

This reaction does require oxygen, so in principle, excluding oxygen from the malaxation process should change this outcome. We have seen several operators attempt this, by flushing the malaxer with an inert gas like nitrogen and sealing the unit. Due to the exposure to air during the milling of the olives, there is enough dissolved oxygen to carry out some reaction with the antioxidants. However, there can be a noticeable increase in the antioxidant content when stirred under nitrogen.[5]

5.3.3.4 Pigments. Looking at different EVOOs, you might notice that the color varies from light orange to golden yellow to bright green. Since TAGs do not absorb visible light, the color must be coming from some other molecules. These pigment molecules (green chlorophylls, yellow lutein, and orange carotenes) start out in the skin and special cells of the olive that are devoted to photosynthesis, not with the TAGs in the vacuoles. As discussed in Chapter 4, the level of the chlorophyll (Chl) pigments in the olive changes during maturation: unripe green olives have a lot of Chl, mixed color olives have less Chl, and mature black olives even less. The presence of carotenes and luteins depends less on the ripeness of the olive. Today, oils are typically made from olives of mixed color (green and deep purple), so the pigments that mix with the TAGs are both chlorophyll and carotene. Note that anthocyanin, the molecule responsible for the purple or black color of a mature olive and discussed in Chapter 4, is polar and does not mix well with the TAGs. It is carried away with the fruit water and contributes to the black color of this byproduct.

Chl and the carotenes (shown here in Figures 5.14 and 5.15) start their journey safely tucked into a membrane containing them. This membrane is like a sandwich, with a nonpolar filling and a polar crust. Because it is rather thin, it is easy to break into little water-soluble pieces during the grinding process. Our pigment molecules will then be floating along inside these pieces, beginning to contact either water or oil as they travel. As the pigments are very much nonpolar, if they contact water, they will not leave the membrane patch. While carotenes are robust and do not react with water, Chl is very likely to lose its one component that is very water-soluble: the magnesium ion which is colored blue in our picture of Chl. Replacing the magnesium ion

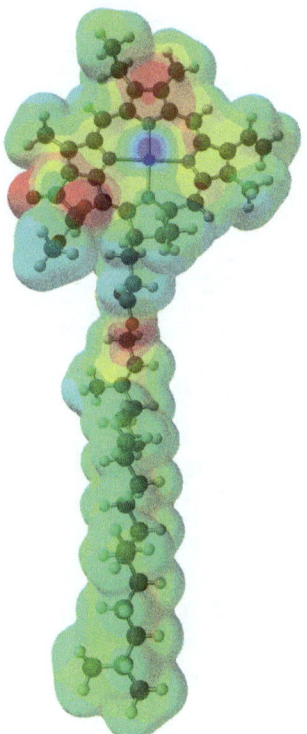

Figure 5.14   Chlorophyll a.

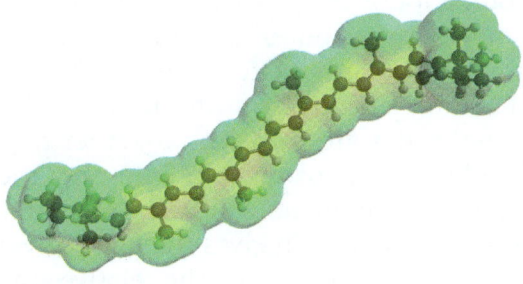

Figure 5.15   β-Carotene.

with a pair of hydrogen ions turns chlorophyll into a new molecule named pheophytin, shown in Figure 5.16. Because pheophytin has very nearly the same light-absorbing structure, it is still green, but has a distinctly darker hue. Water contact during processing turns almost all of the chlorophyll into pheophytin.

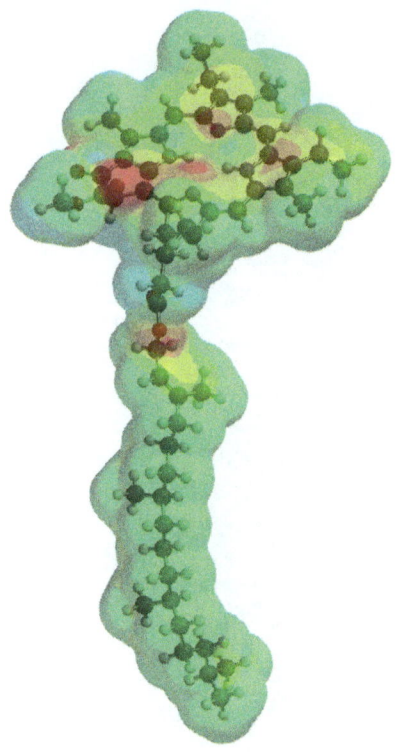

Figure 5.16   Pheophytin.

As it turns out, a lot of the pigment (chlorophyll, pheophytin, or carotenes) never moves out of these membrane patches and is washed out with the olive waste at the end.[6–9] To get into the oil, the patches must encounter a droplet of TAGs. Remember that our TAG molecules are gathering together to make bigger and bigger droplets. If the patch runs into one, there is a very good chance that the pigments will jump ship and transfer to the oil droplet. The pheophytin gives the oil its green color, and the carotenes give it its yellow color. The relative amount of each pigment is different for each type and maturity of olive, giving us a wide range of colors at the end.

5.3.3.5   Steroids and Other Compounds.   There are some steroids in olive oil. Often, the only steroid that people are familiar with is cholesterol, which as an animal steroid, is not present in olive oil or any other plant. The steroids in olive oil are called "phytosteroids," which is a fancy term for "plant steroids." The most common one is β-sitosterol, shown here in Figure 5.17.

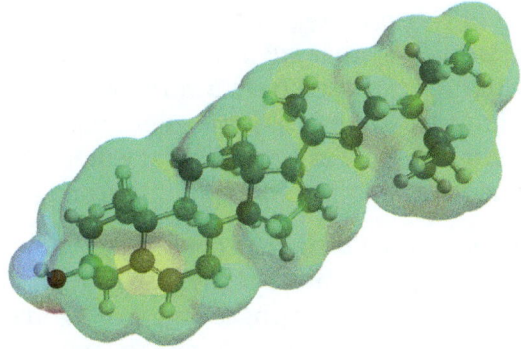

Figure 5.17 β-Sitosterol.

With all the green on the surface, showing its nonpolar nature, you can see why it's soluble in the oil.

These molecules are rather stable, and there are no chemical processes to change them, so they usually pass through malaxation unchanged. The same is true for squalene, a related compound that acts as a chemical precursor to the steroids.

5.3.4 Choosing Sides: Partition Between Oil and Water

As malaxation progresses, the olive paste is coalescing into a material which has more discrete oil and water phases. This forms the stage on which the drama of the molecules' choice is set. These choices are made almost entirely based on the solubility of the materials involved. Offered the choice, a very nonpolar molecule like carotene will almost exclusively dissolve in the oil. Likewise, very polar molecules like oleuropein will choose water. There are, however, many molecules whose allegiance is not so clear. Their story will depend more on the conditions in which the choice is offered.

5.3.4.1 Solubility and Phase Separation. Before we continue with our discussion, we have to make sure that the definitions and issues with solubility are clear. In Chapter 1, we explained that oil and water do not mix because water is polar and oil is not, which is mostly true. Mixing equal quantities of oil and water will very reliably make two liquid phases and we say that they are insoluble. It turns out, however, that with very sensitive instruments, we can discover that a tiny bit of oil can be found in water and a tiny bit of water can be found in the oil, and it might more properly be said that they are mostly insoluble.

Other kinds of compounds, like sugar and water, dissolve very easily and we would say that sugar has a high solubility in water. On the other hand, sugar will have a very low solubility in oil.

There are intermediate cases: molecules such as the phenolics have a moderate solubility in both oil and water. These molecules have an intermediate polarity and are an important part of our story.

5.3.4.2 Choosing Sides: Partition Between Two Phases.   One of the crucial events during malaxation is the distribution of the key molecules between the growing oil phase on the top and the water layer on the bottom. You can probably imagine the natural solubility for a compound in each solvent will be very important in determining how this works. There is a very simple experiment that allows us to figure out what is going to happen. Taking a known amount of a chemical that you are studying, place it in a container with equal quantities of two solvents that don't mix (oil and water). For this to work, it is important to be sure to allow the mixing process all the time it needs so that the mixing is finished and the composition of the two phases has stabilized. Usually the container is shaken to increase the surface area of solvents contacting each other, since that's where any transfer will occur. Once all of the transfer has quieted down, the amounts of chemical in each layer are measured and compared. If the chemical is more soluble in the top oil layer, there will be more in the oil phase. If the chemical is more soluble in the bottom water layer, there will be more in the water phase. If the solubilities are about equal, the amounts in each phase will be equal. Figure 5.18 illustrates these three cases.

How do olive-related compounds fare in this experiment? In Figure 5.19, we show some examples.[10] Note that oleuropein chooses the bottom water phase exclusively and is thus not found in EVOO in significant amounts. Other antioxidants distribute themselves between the water and the oil. Tyrosol is slightly more soluble in the bottom water layer than in the top oil layer. The third compound, DHPEA-EDA, is more soluble in the top oil layer. DHPEA-EDA does not have a simple name like the others but its chemical name is ([2-(3,4-hydroxyphenyl)ethyl(3*S*,4*E*)-4-formyl-3-(2-oxoethyl) hex-4-enoate]). You see why we use the short name DHPEA-EDA. Pheophytin, the main green pigment is found exclusively in the top oil phase, or in its intact membrane patches, but never in water.

The experiment as shown presents a compound with equal amounts of the two solvents. What happens if you increase the

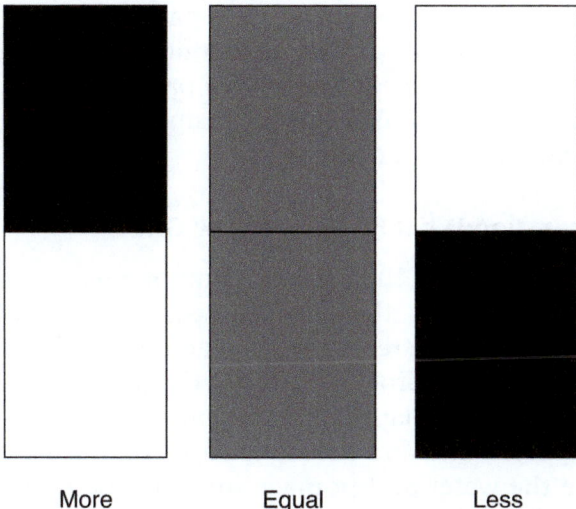

More   Equal   Less

Figure 5.18 Solubility of a compound in the top oil layer *vs.* the bottom water layer. A darker shade means more test compound in that layer.

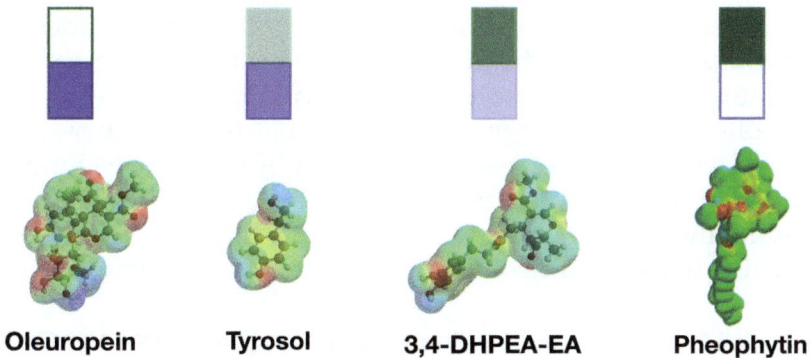

Oleuropein   **Tyrosol**   **3,4-DHPEA-EA**   **Pheophytin**

Figure 5.19 Olive compounds and their partition into the bottom water layer (blue) or top oil layer (green). The intensity of the color shows how much of the substance would partition into that phase.

size of one of the layers? In the extreme cases, it doesn't make much difference. One would still not find oleuropein in the oil or pheophytin in the water. However, for compounds with intermediate solubility, increasing the volume of one of the phases will increase the amount of that compound that will dissolve in the larger phase, naturally reducing the amount in the other. This is why the style of processing can have such an impact on the amounts of antioxidants. Adding more water to the process, whether it is during malaxation or separation, will reduce the amount of antioxidants in the oil. Controlling the amount of water added to the process is a critical part of producing a good oil.[11] Fortunately, there is a relatively simple test that the plant manager can do: taste the oil!

5.3.5 Separation:What Ends Up in the Oil?

The concentrations of the various components are established during malaxation by the complex combination of physical changes in the olive paste, chemical reactions, and solubility distribution. When that process is done, the paste is pumped into a decanter. This is a relatively simple physical process: the oil is less dense than water and moves towards the inside of a centrifuge, while the water and pomace move to the outside. All the changes produced by malaxation are captured in this separation, and the only remaining steps are to clean up the oil for long-term storage. Even a high-quality machine will leave just a bit of the water phase behind, perhaps held by the few molecules left that can act as surfactants. This water phase contains much more bitter compounds than the oil and can also contain enzymes that degrade the TAGs and antioxidants. It will require at least one of the following steps to become ready for bottling: a higher-speed centrifuge will remove more of the water by the same technique, filters will remove water by attraction to the filter media (made essentially of paper), or racking will allow the water to coalesce and sink to the bottom of the tank, where it can be drawn off.

5.4   WHY PRODUCE OLIVE OIL THAT IS NOT EXTRA VIRGIN?

With the focus on extra virgin oils that we have been following, you might wonder why anyone would produce anything else. After all, the best price per liter comes from EVOO, which would seem to lead to greatest profit. However, if it is not possible to make EVOO, for a variety of reasons, it might still be possible to produce virgin olive oil, oil for non food uses called lampante oil, or to send the oil to a processing plant to be chemically refined and sold as Olive Oil.

First, if the olives are not of high enough quality, then nothing in the processing can save that. Those olives still have some worth, and the producers who have picked them want to capture that value. Owners of a family grove who do not have the resources to harvest otherwise can lay down nets at the beginning of the harvest season and collect olives every few weeks. Of course, these olives will have been on the ground, exposed to the elements, and sitting in the sun for that time, and will be of poor quality. However, some oil may be better than no oil at all from the trees. Others are forced into making it due to circumstances involving pest infestation, poor weather at harvest time, or the inability to keep up with the harvest (over-ripe olives can produce lower quality oil). Similarly, if the olives started off in good shape, but were treated badly on the way to processing (see in Figure 5.20 the sitting of olives in a storage yard for a day or more), they will not produce good oil.

Finally, many producers, looking at a market that may not give very different prices for different oils, might decide to go for yield instead of quality. One common tactic in certain parts of the world is to run the malaxation and separation at an elevated temperature (we've seen about 40 °C several times). This gives better quantities, but the oil can have a softer quality without much aroma, and can smell "cooked." If this matches local tastes, that can be the best solution, especially if the olives are a bit marginal.

Figure 5.20 Olives moldering in the sunshine outside an olive press. Notice black mold on burlap sacks.

5.4.1   What Quality Oil is Actually Produced in the World?

It's hard to get more than an indication, but using data from the International Olive Council, the Mediterranean production is a bit over 50% EVOO, over 20% virgin olive oil, and as much as 20% lampante olive oil. There is considerable variation from country to country in this value. In the latest reports of quality available, Israel reported 90% extra virgin oil production, while Morocco reported 5% (leading to a strong initiative to increase the quality). There is a trend towards increased production of EVOO, but this typically requires the replacement of production facilities and change in harvest practices, which are not affordable in developing economies and slow to come in developed ones. Mostly because of the age of the general industry, the quality in the "New World" production is much higher, typically over 90% EVOO.

Lampante oil is not sanctioned by many governments as fit for consumption. Consequently, it must either be used for lamps (not very common during the last century) or chemically transformed for other uses. This transformation is called "refining." The product, called "refined oil" is not of higher quality than EVOO, but it is arguably more useful than lampante. Hence, "refined" is a relative term, and one should remember the source material.

5.5   MODERN REFINING

The process of refining produces an oil that is edible but has also lost much of its character. Even for chemists, it would be difficult to tell the difference between refined olive oil and oil refined from hazelnuts or avocados. What's left after the treatment is an almost pure triglyceride mixture with the pigments, aromas, phenolics, and other small molecules removed.

5.5.1   Treatment with Base

Usually, the first step in refining is the addition of a calculated amount of a base, such as caustic or lye (sodium hydroxide or potassium hydroxide) to the oil. Heating this mixture causes these chemicals to react and the free fatty acids are converted from nonpolar molecules into polar salts of a special type called soaps. These are the same kind of molecules as those found in traditional soap, and the refiners often collect them for that use.

The key is that, as salts, they are quite soluble in water and can be washed away from the oil.

Note that any phenolic compound present will also react with the base in the same way to make a water-soluble salt. After this step, there will be no phenolics left in the oil.

Finally, water is added and the two phases are separated using a decanter. The oil produced will have a very low free-fatty acid content, typically below 0.1%, and no phenolics.

5.5.2 Decolorizing

After this first step, the oil can be quite deeply colored, and certainly not with a color that is associated with olive oil. Residual pigments will have browned, and the cooking process will have produced new brown or black pigments. These can be removed by the use of a mineral called bentonite, which is also used to clarify wines. When ground into very fine particles, bentonite has a very large surface area and is attracted to materials that are at least a little bit polar. This includes all of the pigments and any residual salts and metal ions from the previous step. If there is a concern that some other toxic compounds have crept in (polycyclic aromatic hydrocarbons, or PAHs, are a common byproduct of pomace oil production), activated carbon or charcoal may be added.

This step is usually carried out by stirring the slurry at about 100 °C for a half hour or so, then cooling a bit and filtering the result. At this point it might be essentially colorless. Any residual antioxidants, such as vitamin E, will be removed at this step as well.

5.5.3 Vacuum Distilling/Stripping

While the oil may look good, it will probably not smell very good here. Breakdown products from the previous high-temperature processes, along with off-aromas from the original oil will still be present. The final step removes these. Since aromas by nature must be somewhat volatile, it makes sense to remove them by heating the oil under vacuum. The heat and reduced pressure make the move from dissolved in the oil to floating in the atmosphere above it attractive to all aromas, and they can be removed. Just to be sure they make it out of the bulk oil, water is added to the oil (held at low pressure and over 200 °C). The water instantly vaporizes, adding to the vapor flowing out of the container and enhancing the removal of the aroma molecules. This may take an hour or so to complete.

5.5.4   What Could Possibly Go Wrong?

Think about how long the oil spends at high temperature in the above process. The TAGs in olive oil are quite resistant to change, even at these temperatures, so they should be more or less intact. There is some concern that trans-fats will accumulate in the oil, simply due to the high temperature.[12] There are quite a few new compounds formed during this process, few of which have been identified. While they are probably innocuous or removed by the refining process, the complete story is not known.

5.5.5   Is Refined Olive Oil Any Good?

This is a question of preference. Let's remember first that the triglyceride mixture should be essentially unchanged from the original oil. This is a very good thing, because it is healthier than many other oils or fats and very stable to decay. Refined oil can also be a very good choice for high temperature cooking (which would destroy many of the qualities of EVOO) or for mixing with other oils where the flavor should not come from the oil itself. It will not, however, add any flavor to your food or yield many of the health benefits of EVOO that come from the phenolics and other compounds that have been purified away. Confusion over market labels for olive oils is understandable. Words like "pure," "light" and "refined" used on a label convey a sense of higher quality or increased health benefits when in fact they are the exact opposite.

5.5.6   Pomace Oil

Crude pomace oil is extracted from the olive waste by the use of solvents such as hexane, which dissolve the oil and allow an easy separation from the waste. After the extraction, most of the solvent is removed. Crude pomace oil is not considered food grade, but may still be useful for industrial purposes, such as soap-making.

Chemical extraction of olive oil using hexane was developed in the mid 1900s and considered an improvement over old-fashioned separation methods. The extraction process was heralded as being more efficient at separating the oil and far cheaper, as shown in the Shell advertisement from 1948 (Figure 5.21). Since the boiling point of hexane is much lower than the oil, heating the mixture should drive off the hexane and minimize the level

Like other sources of edible oils and fats, olives now play in "the Majors." Every drop of olive oil is in demand. How many drops can an olive produce?

That chemical symbol making the put-out represents a Hexane—which extracts more oil from the olive. When Shell scientists first got Hexane from petroleum, there was little reason to think that as an "extraction solvent" it might add indirectly to the food supply ...

But that day has arrived. Shell is the principal supplier of Hexanes for olive oil extraction. The first "take" of oil comes from pressing the olives – it must be done gently, to avoid overheating and turning the oil rancid. The crushed fruit and seeds, still rich in oil, are then treated with Hexane. Its nimble chemical fingers extract the remaining oil – the oil is really OUT.

The low, narrow boiling range of Hexane is the key to the process. After extracting the oil, the solvent is boiled off, at comparatively low temperature, and used again and again. It's so volatile that not a trace of odor or taste remains in the oil.

Other agricultural by-products – grape seeds, rice bran ... tomato, flax, sunflower, and peach seeds ... are ground up and given the "Hexane treatment" – producing a number of valuable materials for industry.

Production of Hexanes from petroleum is only one achievement by which Shell demonstrates leadership in the petroleum industry, and is petroleum products. Wherever you see the Shell name and trade mark, Shell Research is your guarantee of quality.

Figure 5.21   Shell Oil's advertisement from July 5, 1948, showing the olive oil refinement process. Shell's byline "horizons widen through shell research" highlights the appeal at that time of getting rid of olive

of any residue in the refined oil product. Consumers at the time appreciated refined olive oil for its clarity and lack of odor or taste. Today, most soybean oils, rapeseed oils, and sunflower oils are still processed by hexane extraction. But tastes change, and many of today's consumers enjoy the taste and aroma of a good EVOO. Other consumers prefer to minimize their exposure to chemical solvents like hexane.

5.6 CONCLUSION

It is quite impressive how much goes on in the hour or so between the olive's arrival at the press and the departure of the oil from the end of the process. During that hour, the plant manager makes many decisions about the processing of the olives that will determine the quality of the oil. Some of these important decisions are how to best prepare the olives for crushing, whether or not to add talc or enzymes, how fast and how finely to pulverize the olives, how long and at what temperature to malax the fruit, how much water to add to the decanter and finishing centrifuges, and whether or not to filter the oil. From the smallest olive presses that process 1 kg per hour to large industrial presses that process up to 100 tons per hour, one characteristic that unites all the managers we have met is a love for what they do and a passion for doing it right.

---

odors and flavors and presenting consumers with the "pure" hexane extracted oil. Copyright Shell Brands International AG. The text in the advertisement reads "Like other sources of edible oils and fats, olives now play in "the Majors." Every drop of olive oil is in demand. How many drops can an olive produce? The chemical symbol making the put-out represents a hexane – which extracts more oil from the olive. When Shell scientists first got hexane from petroleum, there was little reason to think that as an extraction solvent it might add indirectly to the food supply. But that day has arrived! Shell is the principal supplier of hexane for olive extraction. The first press of oil comes from the olives – it must be done gently, to avoid overheating and turning the oil rancid. The crushed fruit and seeds, still rich in oil, are then treated with Hexane. Its nimble chemical fingers extract the oil — the oil is really OUT. The low narrow boiling range of hexane is the key to the process. After extracting the oil, the solvent is boiled off, at comparatively low temperature, and used again and again. It's so volatile that not a trace of odor or taste remains in the oil." The corporate logo reads "Horizons Widen through Shell Research".

REFERENCES

1. A. Bakhouche, J. Lozano-Sánchez, C. A. Ballus, A. Bendini, T. Gallina-Toschi, A. Fernández-Gutiérrez and A. Segura-Carretero, *Talanta*, 2014, **127**, 18.
2. A. Ranalli and G. De Mattia, *J. Am. Oil Chem. Soc.*, 1997, **74**, 1105.
3. A. Ranalli, T. Gomes, D. Delcuratolo, S. Contento and L. Lucera, *J. Agric. Food Chem.*, 2003, **51**, 2597.
4. L. S. Artajo, M. P. Romero and M. J. Motilva, *J. Sci. Food Agric.*, 2006, **86**, 518.
5. A. Yorulmaz, A. Tekin and S. Turan, *Eur. J. Lipid Sci. Technol.*, 2011, **113**, 637.
6. A. Ranalli, P. Cabras, E. Iannucci and S. Contento, *Food Chem.*, 2001, **73**, 445.
7. M.-N. Criado, M.-P. Romero, M. Casanovas and M.-J. Motilva, *Food Chem.*, 2008, **110**, 873.
8. M. J. Moyano, F. J. Heredia and A. J. Meléndex-Martinez, *Compr. Rev. Food Sci. Food Saf.*, 2010, **9**, 278.
9. A. Giuliani, L. Cerretani and A. Cichelli, *Crit. Rev. Food Sci. Nutr.*, 2011, **51**, 678.
10. P. S. Rodis, V. T. Karathanos and A. Mantzavinou, *J. Agric. Food Chem.*, 2002, **50**, 596.
11. M. Servili, R. Selvaggini, S. Esposto, A. Taticchi, G. Montedoro and G. Morozzi, *J. Chromatogr. A*, 2004, **1054**, 113.
12. *Heat Induced Cis/Trans Isomerization in Vegetable Oils and Oleic Acid*, http://blogs.longwood.edu/incite/2012/01/30/heat-induced-cistrans-isomerization-in-vegetable-oils-and-oleic-acid/, accessed April 2015.

CHAPTER 6

Delivering Quality and Assuring Authenticity

The frenetic work of processing is done, which also means that the harvest season is over. Producers begin to get more sleep and reacquaint themselves with their families. In most places, the oil is safely resting in stainless steel tanks, blanketed in inert gas and releasing the last tiny bits of moisture. The next important step is to deliver the oil to its important destination: the consumer.

According to the International Olive Council (IOC), a little over three million tonnes of olive oil (tonne = 1000 kg) were consumed in the world in 2014. That's enough to fill over 1300 Olympic-sized swimming pools (which is a really bad idea for storage, as you'll see later). Where in the world does all this oil come from, and how does it get to you?

Even with a fairly substantial increase in production outside of the Mediterranean, the bulk of the world's olive oil still comes from within that region. This oil is shipped all over the world, in bulk or in consumer packaging. A fairly large reserve of oil is maintained in the major producer countries, for internal consumption and as a hedge against a bad production year. Each producer will decide which oil to sell at home and which to export.

The Chemical Story of Olive Oil: From Grove to Table
By Richard Blatchly, Zeynep Delen Nircan and Patricia O'Hara
© Richard Blatchly, Zeynep Delen Nircan and Patricia O'Hara, 2017
Published by the Royal Society of Chemistry, www.rsc.org

In addition, oil may be exported to a second country, blended with other oils, and then re-exported to a third country.

In this chapter, we will explore the surprisingly complex process of getting the oil out of the tanks and into packages and ultimately to your kitchen. We will also look at the many ways in which oil is evaluated for quality and authenticity. This last topic is of increasing importance as the global market for olive oil expands into new territories.

6.1 PACKAGING

When the oil is removed from the storage tanks, it is often transferred directly into consumer packaging. In addition to identifying the oil, indicating the contents, and producing an attractive product, packaging should also protect the oil. Olive oil is sensitive to oxygen, and even a little bit of oxygen under the wrong conditions can do a lot of damage. The other enemy of olive oil is light. Any light which excites the chlorophyll derivatives in the oil may activate the oxygen and accelerate destruction of the oil. It doesn't take much light to cause trouble in the presence of oxygen. Therefore, the packaging material should be impervious to oxygen and should block as much light as possible.

The best solution to this problem is to avoid both oxygen and light. Many producers package the oil under an inert gas in metal or other opaque containers (including multilayer foil bags). While this hampers visual examination of the oil, keeping it away from light is undeniably better for storage. A properly packed olive oil should be resistant to decay while unopened.[1] Even so, it is still a fruit juice, and will naturally decline in quality as it sits. Packaging should also protect the oil even after the oil is opened by the consumer.[2]

6.1.1 Inert Atmosphere Flush

Because oxygen is such a problem, any modern bottling process includes a step to replace the air in containers with an inert gas such as nitrogen or argon. Flushing with one of these inert gases ['gasses' is the present verb] eliminates the oxygen in the container. These gases are chemically unreactive and odorless, and are also used to flush the tanks in long-term storage. This does add a bit to the cost of the oil, but extends the shelf life by slowing the breakdown of the oil.

6.1.2   Container Material

Due to the many functions of a container, the choice of material can be difficult. If it is important to see the oil, then glass should be used. However, glass is subject to breakage and is relatively heavy, and even if colored, it may let light into the oil. Colored glass bottles reduce light intensity just as sunscreen reduces light damage to your skin, but in the same way considerable damage can still be done when sun or light exposure is for extended periods of time. The damage caused by light is accelerated when the bottle is opened, as oxygen enters the container when opened (with the possible exception of a collapsible container). Labels that cover most of the surface of the bottle may reduce light intensity sufficiently to add to the shelf life of an opened bottle of olive oil.[2]

Metal containers are traditional, are a good choice as they are impervious to oxygen and light, and are strong and unbreakable. Like glass bottles, they allow air to enter when the oil is being used, which may accelerate aging. Metal may also react with any moisture left in the oil, causing corrosion or accelerating oil breakdown. Due to the tendency of the crimped areas to collect solids from the oil, metal containers should not be re-used.

A relatively recent addition to the array of containers is a treated, multilayer plastic pouch contained in a cardboard box. Commonly known as a "bag in a box," this has been successfully used for wine and juices, and works quite well for olive oil. These plastic pouches are usually made of oriented polyethylene terephthalate (PET) and are shiny due to the thin coat of metal deposited between the plastic sheets. They are resistant to oxygen, impervious to light due to the metal layer, extremely strong, and very light. When the oil is drained from the pouch, it collapses, meaning that no air enters the container. Their only drawback to seems to be the difficulty some users have in deploying and using the spout, which must be toward the bottom of the container rather than on the top. On the other hand, these are the lightest to ship (and provide efficient delivery of empty containers to the bottling facility).

Another relatively recent innovation is the single-serving pouch of olive oil, as shown in Figure 6.1. Made of a foil pouch lined with food-grade plastic, this can be snapped open and poured directly on the food.

The type of plastic bottle used for soft drinks is very permeable to oxygen, however, and is not suitable for long-term storage of extra virgin olive oil. Its low cost leads to its use in many informal markets, however, and it might be acceptable for oil that will be used within a week or two.

6.1.2.1 Closures. A noted weak point of most of these containers is the closure. The pop-up screw top, or standard screw top on a bottle or can, or the spout on a collapsible bag must be securely attached to the container. Consumers are acutely aware of whether the closure allows the oil to flow easily and avoid dripping.

An additional factor is important for containers that go to restaurants. Some restaurants try to cut costs by refilling olive oil bottles daily, rather than providing new ones. While this may sound like an environmentally friendly thing to do, if oil is refilled, the residual oil can be very old, and has been exposed to air for a long time. Rancid oil reflects badly on the producer if their label is on the oil. We have seen tamper-proof tops to

Figure 6.1 Single serving packet of olive oil. Used with permission from http://packitgourmet.com.

prevent restaurants from "topping up" the bottles. The European Union has recommended that refillable containers be banned in restaurants (EU regulation 29/2012). Although that provision of the Regulation is merely advisory, it has generated considerable anger in several EU member states who see this as unwarranted micromanagement and legislative over-reach.

6.1.2.2   What Can Go Wrong During Shipping and Handling?.   The worst enemies of olive oil are oxygen and light.[3] Oxygen attacks many of the important molecules in olive oil, including tri-glycerides, as we discuss later in this chapter. We have found that almost all oil has some relatives of chlorophyll (usually pheophytin), even oils that don't look green. Pheophytin absorbs light very well, producing an excited pheophytin molecule. In the absence of oxygen, this is not a big problem, as the excited molecule just loses its energy and returns to its original state. However, if oxygen is present, an excited pheophytin molecule can transfer this energy to the oxygen, producing something called "singlet oxygen." As damaging as oxygen can be in long-term contact with the olive oil, singlet oxygen is many times worse. If given a chance, it will react with the triglycerides, especially the polyunsaturated fatty acids (PUFAs), and break them down. This can lead to accelerated aging and rancidity. With some luck, the singlet oxygen will encounter an antioxidant molecule before it encounters the triglycerides, reacting and neutralizing both the singlet oxygen and the antioxidant molecule. This reduces antioxidant content but preserves the rest of the oil, including the aroma. Remember, though, that the number of triglyceride molecules is much greater than that of the antioxidants.

It's worth mentioning that whatever challenges an oil may face, elevated temperature will make them worse. All degradation reactions run faster at elevated temperatures (a common rule of thumb suggests that reactions double in rate for every 10 °C increase in temperature). Storage in a relatively cool location is best at all parts of the supply chain.

6.2   THE ROLE OF OXYGEN AND LIGHT IN OIL DECAY

We tend to think of oxygen as a completely necessary and helpful substance. Certainly, none of us can live without it for more than a few minutes. However, it also poses challenges to humans and olive oil alike, slowly reacting and resulting in aging or other oxidative processes.

There are several forms of oxygen that can cause damage. The form of molecular oxygen we breathe is the least dangerous, but is the source of oxygen atoms for the rest. Even in its most benign form, given enough time an oxygen molecule can react with a molecule of fatty acid, for example, to remove one of the hydrogen atoms from that molecule. This creates a kind of raw end, bearing an unshared electron (a "radical"), which must react with something else. Using another oxygen molecule, the combination can attack a second fatty acid molecule to create an organic peroxide and a new radical on the second molecule. The most susceptible sites in the molecules are the positions next to the double bonds, even more so if there are two double bonds flanking the site. This makes polyunsaturated fats (PUFAs) most reactive with oxygen, monounsaturated (MUFA) much less so, and saturated fats least reactive.

Figure 6.2 shows a model for the reaction for a typical PUFA (such as linoleic acid, shown at the left of the figure). The dark electron clouds near positions 9 and 13 mark the position of the double bonds, and the extremely reactive position 11 is in the middle. Each dark cloud is a portion of the electrons associated with a double bond (unsaturation), which is composed of a core pair of electrons (not shown) and a looser pair of electrons above and below the bond. These loose electrons can interact with neighbors in interesting ways. Their presence makes it easier to remove the hydrogen from position 11, as the unshared electron

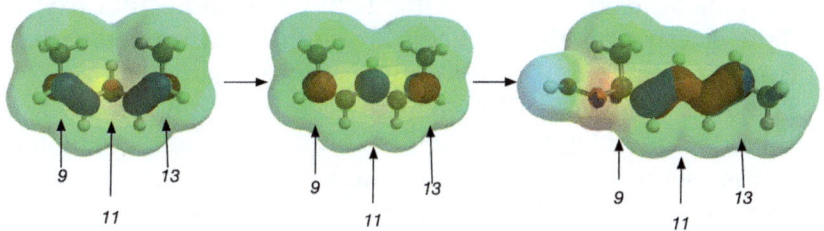

Figure 6.2   Development of peroxides in oil. A polyunsaturated fatty acid can lose a hydrogen atom to produce the highly unstable radical shown in the middle. This radical can react with an oxygen molecule and a second fatty acid to produce the peroxide shown on the right.

that is generated will mix with the two loose pairs of electrons beside it, spreading out across the five atoms.

The center structure represents this radical formed by removing a single hydrogen atom from the original molecule. The red and blue electron clouds on atoms 9, 11, and 13 represent spreading out of the electrons along these atoms. Since atoms 9 and 13 now have some radical characteristics, oxygen can be added at either end (9 or 13). The peroxide product shown at the right gives the form in which oxygen has been added to the 9-position and is functionally the same as the alternative. The double bonds have returned (on atoms 10 through 13), but the loose electrons are now spread evenly across all four atoms, instead of being isolated from each other. Because this arrangement of loose electrons (called "conjugated") is more stable, it is formed very reliably when peroxides are formed from PUFA. Since the conjugated electrons respond to light differently than the isolated ones, we'll use that quality to test for this fatty acid damage using ultraviolet light.

Because a new radical is formed from a second fatty acid, this cycle can continue for many rounds (even hundreds or thousands) until a chance encounter with another special radical (called an antioxidant; more on that will come at the end of this chapter) terminates the cycle. This is often referred to as a chain reaction and only requires one initiating oxygen to get the chain started. This makes oxygen an especially powerful foe of quality olive oil.[1]

Why is light a bad combination with oxygen? As bad as oxygen seems in this example, the reaction is very slow to start. However, light, when absorbed by a chlorophyll or pheophytin molecule, can activate oxygen into a supercharged form called "singlet oxygen." This molecule has the same formula as molecular oxygen, but the electrons in the molecule are slightly rearranged to make the new molecule thousands of times more reactive.[3]

The bottom line regarding reactions with oxygen is that the oil will be degraded significantly if exposed. This damage includes the loss of positive aroma notes and the generation of aroma defects. It also includes the destruction of antioxidants, which have positive flavor and health effects. Even the pigments can be altered, typically by bleaching a green oil to light yellow or colorless. In the absence of light, this can be relatively slow (measured in weeks, not hours), but in the presence of light, the damage can be done in a few hours.

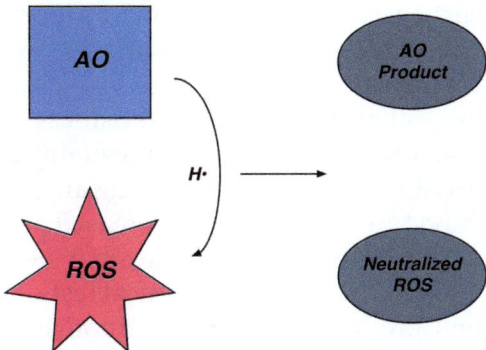

Figure 6.3 The role of antioxidants in destroying oxidants.

6.2.1 The Role of Antioxidants

The secret (well, ok, not so secret) weapon against these reactions with oxygen lies with the antioxidant molecules. These include some vitamin E, but are largely the derivatives of hydroxytyrosol that we call phenolics, as discussed in Chapter 3. These molecules can interrupt the oxidation process by reacting with the radicals as they are formed, making them whole and non-reactive again. While this doesn't return the molecules to their original state, it does prevent further damage. This reaction is a sacrificial one and requires one molecule of antioxidant for every radical neutralized, as shown schematically in Figure 6.3. The antioxidant ("AO") can transfer a portion of its molecule to the reactive oxygen species ("ROS"), which can be molecular oxygen, a peroxide, or a radical. Olive oil antioxidants typically react by transferring a hydrogen atom to the ROS. This reaction neutralizes the oxidant, making it no longer reactive. However, it also consumes the antioxidant, meaning that the supply of antioxidant has been reduced. Obviously, this can only persist as long as the antioxidant supply lasts. Furthermore, as the antioxidant molecules are responsible for part of the flavor, that will also change if enough molecules are consumed.

6.3 WHAT CAN HUMANS DO TO THE OIL?

The production and sale of olive oil is increasingly under scrutiny and subject to governmental and societal regulation. This is generally helpful, as very high quality extra virgin olive oil can be made and delivered to consumers by reputable producers and merchants. Nearly everyone we have met around the world has a desire to deliver the highest quality oil that stems from their passion for their product.

However, olive oil production is a business and is run by humans. The usual human foibles will inevitably come into play: fear, greed, inattention, and ignorance being among them. Fear of losing a year's investment or greed for increased profits may tempt a producer to blend lower quality oils in such a way that enables the oil to pass some of the simpler chemical tests. Oils which were extra virgin when produced can decay unnoticed if they are not tested frequently, and reach the consumer in a rancid condition. Perhaps a producer truly believes that the oil is extra virgin, but has not been sufficiently trained to know the difference.

The classic story of the darker side of olive oil production is told in Tom Mueller's book "Extra Virginity."[4] In the years following publication of this work, prompted no doubt by his vivid description of both good and bad producers, sampling of olive oil began to be carried out by national trade associations and news outlets. High profile stories in Australia, Italy, the United States, and other countries document bottles of olive oil pulled from supermarket or specialty stores, tested, and found not to match their labels in alarming numbers. These stories can involve small producers as well as some of the biggest suppliers in the business.

Bertolli and Bertolucci, two of the largest Italian olive oil brands sold in the US, were recently sued in California for selling extra virgin olive oil in clear bottles that was rancid, and the products were removed from the shelves. It is likely the deterioration of the oil was accelerated by the clear glass of the container. There was a similar case in Italy in 2015. The good news is that these companies have responded to the challenge, and have changed packaging, increased vigilance and standards for chemical testing, and added traceability for each bottle of oil.

6.4   WHAT DOES "AUTHENTIC" MEAN FOR OLIVE OIL?

As we have traveled around the world to different olive-growing regions, this has been one of the more interesting discussions we have participated in. There are many opinions about what "authentic" means. Reports in the news media about various kinds of fraud and deception have raised the public profile of this question considerably. There is a wide variety of tests available to clarify these issues.[5] Common tests for the different grades of olive oil exist to determine free fatty acids (FFA) and peroxides, and to provide "organoleptic analysis." The word "organoleptic" is constructed by considering the sensory "organs" and what they will analyze, or take in (Greek: *Leptikos*). In practice, this analysis is done by a tasting panel and will be described more fully in Chapter 7. However, an oil can be ascribed as having positive qualities according to organoleptic analysis (like grassy or fruity aroma, bitterness, or pungency) or negative qualities (like musty, fusty, rancid, or spoiled aromas). Finally, there is a commonly required measure, somewhat related to damage by oxygen, called the peroxide value. You may have seen some of these reported on bottle labels and wondered what they mean. We will be explaining these, and many more, in this chapter.

For our purposes, the question of authenticity can be divided into four more focused categories, all of which affect consumer satisfaction and the underlying chemistry.

6.4.1 Is This What I Thought I Was Getting?

The first part of defining an authentic product is making sure we are all speaking the same language. One barrier can be found on the labeling on an olive oil bottle, which may imply certain qualities without actually saying that they are present. Strictly speaking, the oil may be "authentic," while the label is misleading in the same way that one would be disappointed after buying a bottle labeled "fresh orange juice" only to discover later that it was not the fresh squeezed orange juice you expected but was made from concentrate.

To be labeled "extra virgin olive oil" a product must comply with strict legislative standards. Perhaps you've seen bottles marked "pure olive oil" or "extra light" olive oil and believed them to be superior products. Hopefully, after you read the section below on what you can learn from the label, you will have changed your mind. After a while, it will be second nature to you to check for these disclosures. These are not necessarily bad products, but they are just not the extra virgin olive oil you were led to believe you were buying.

6.4.2 Is This Olive Oil or Some Other Oil?

It may seem logical that a bottle labeled "olive oil" should contain only oil from olives. However, a small percentage of the oil in circulation may contain oil from other sources. The usual contaminant oils may be hazelnut, sunflower, canola, or partially hydrogenated seed oils. When not disclosed on the label, this is clearly fraudulent activity, and stems from a supplier simply wanting to make more money (by inflating the volume sold), or a supplier stuck in a contract to supply a volume of oil that is not available this year. In addition to being fraudulent, this is dangerous to those who may be allergic to the contaminants, and the oil clearly will not give the same health benefits promised for good quality olive oil.

6.4.3   Does the Quality Match the Label?

This is a more subtle question, which requires knowing a bit about what labels say. Since our focus is on extra virgin olive oil – words that describe the highest grade of olive oil – let's start there. In most countries, to be labeled "extra virgin olive oil" the olive oil must be free from defects in the aroma, have at least one positive trait (such as fruitiness or bitterness), and have a free fatty acid content of 0.8% or lower. Oil that is produced by inferior olives or an inferior process will not meet those standards. To label such oil "extra virgin olive oil" is dishonest and illegal, whether this is a result of conflicting standards (just how strong must a defect be before it can be detected by a sensory panel?) or from deliberate mislabeling. Another possibility is that the oil started out extra virgin, but problems encountered during the shipping or storage, such as high temperatures or exposure to light, caused the oil to deteriorate to the point where it is no longer extra virgin. This is not unusual and will be explored in greater detail below. Probably the most common problem with supermarket or restaurant extra virgin olive oil is that the oil is just too old and has become rancid. Rancid oil will not hurt you, but it will not have the flavor, aroma, or health giving properties of extra virgin. Always check the label for a best buy date or, even better, a bottled date and make sure you are getting oil as close to its harvest date as possible.

6.4.4   Where Does the Olive Oil Come from?

At a basic level, this question may seem less important. After all, if an oil is properly labeled as "extra virgin olive oil," why should we care where it comes from? The issue is partly a question of "truth in advertising": a bottle labeled "produced in Italy" should contain oil from olives produced in Italy. Serious students of olive oil can sense the taste differences among oils produced in different regions, particularly if the oil is to be used in a particular dish.

This is complicated by the fact that the word "origin" may have several meanings. It might mean where the olives are grown, where the oil was bottled, to or where it was labeled and exported from. Traditionally, these three locations were the same (in fact, often in the same building), but increasingly this is not the case. Olives grown in Spain can be pressed in Spain, stored in stainless steel tanks and shipped to Morocco, filtered and bottled there, and shipped to Italy for labeling and distribution. It's important to learn to read labels for what they do and don't tell you.

6.5   HOW DO CHEMISTS KNOW IT IS OLIVE OIL?

As mentioned above, when we buy a bottle of olive oil, we are relying on the production chain to supply a product with the characteristics claimed on the label. There are many opportunities along the chain for individuals to cut corners, to willfully add less expensive products to dilute the olive oil. We are fortunate that most producers are honest and careful. This process is now monitored in many countries, and we are beginning to develop some useful tools for keeping track of the results.

Representing a product labeled "olive oil," when it contains something other than olives is fraudulent. One factor in favor of catching those who engage in such activities (they are not "criminals") is that relatively simple accounting techniques can be an early warning system triggering a more thorough analysis. If a facility takes in 1000 liters of olive oil and puts out 3000 one-liter bottles of product, it is clear that they have misled the consumer. Of course paperwork can be fabricated, especially when the product crosses several international borders. To catch the culprits, we need more sophisticated tools.

6.5.1   What is Found in Other Oils?

To explore what tools might be available, let's look at the chemical fingerprints of some oils commonly used to dilute olive oil (hazelnut, refined olive, sunflower, canola, partially-hydrogenated flower seed oils). Each oil has its own fatty acid profile that serves as a unique chemical fingerprint that can be used in much the same way as a detective uses a person's fingerprint to deduce their presence at a crime scene.

The origin of the mixture of fatty acids in olive oil is developed in Chapter 4. As shown in Figure 6.4, most seed and nut oils, with the exception of hazelnut oil, have a noticeably different fatty acid composition from olive oil. They are much lower in oleic acid, and often much higher in the PUFAs. For the presence of oils like canola and safflower, that alone can be enough to show that the oil is contaminated. There is a component in refined olive oil and in partially hydrogenated seed oils that signals their presence: *trans*-fatty acids. These will be explained in more detail later in this chapter. Since there are no *trans*-fatty acids in extra virgin olive oil, the presence of any significant quantity is a clear sign of tampering.

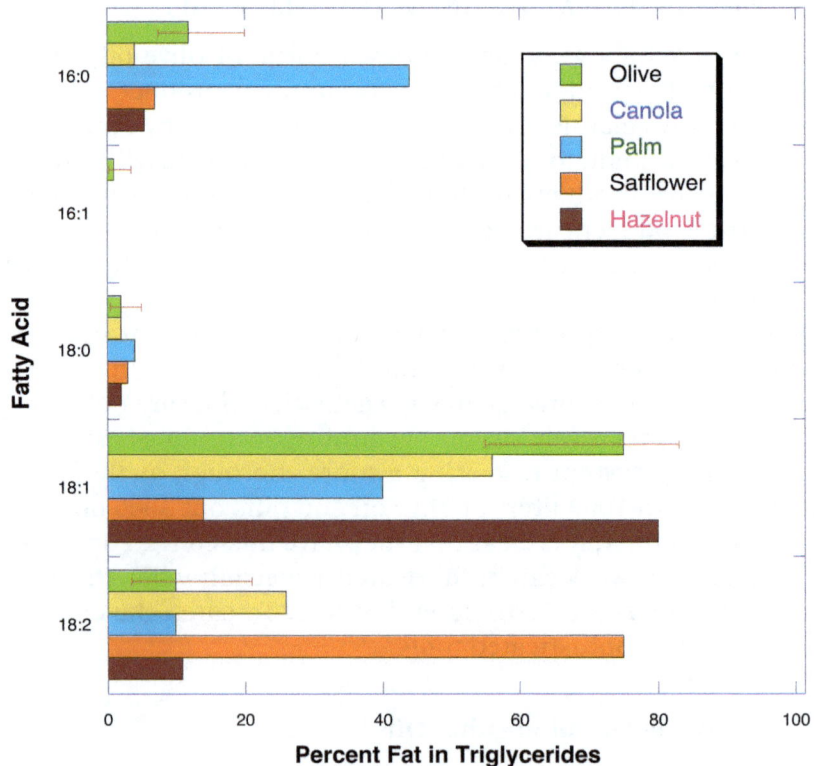

Figure 6.4   Mixture of fatty acids found in olive oil and other common seed and nut oils. The bars show the average percentage of each fatty acid in the oil. For olive oil, the allowable range is shown with the range bars extending to either side of the average.

Hazelnut oil is a different story. Its fatty acid profile almost perfectly matches that of olive oil: high in oleic acid, low in PUFA, and low in saturated fats. This makes hazelnuts a delicious and healthful snack, and hazelnut oil a tempting additive to olive oil. At times when the price of this nut oil is lower than olive oil, it is tempting to substitute one for the other. To do so without disclosing the presence of nut products on the label could have devastating consequences if an individual allergic to nuts consumes olive oil diluted with hazelnut oil. To prove the presence of hazelnut oil, we need another tool.

When the fatty acid profile fails, another valuable tool is the analysis of the steroids in the oil. Steroids are a class of biomolecules that serve as hormones, anti-inflammatory agents, vitamins, and many other important functions. Because we can't taste them, not much has been written about steroids in oils. However, there are thousands of different steroids in plants and animals, and the mixture in olives is unique. The presence of a steroid in the oil that shouldn't be there is another clue, like a cigarette butt in the room of a non-smoker.

6.5.2 Best Tests for Identity: Fatty Acid Analysis

The analysis of fatty acids has two steps that help us reduce some of the complexity of triacylglycerides (TAGs) and make the measurement easier. Think about all the different combinations of outfits you could make if you had 10 shirts, 10 pairs of trousers and 10 pairs of shoes: at least 10*10*10, or 1000 different outfits! It is a lot easier to describe what's in the closet instead of cataloging all the outfits.

In a similar way, TAGs are most commonly analyzed by a chemical process in which the first step is breaking down the complex TAG molecule into free fatty acids and glycerol, converting the free fatty acids to fatty acid methyl esters (known as FAME) and finally analyzing those individually by an analytical method known as gas chromatography, described briefly in Figure 6.5. Overall, the process is quite reliable and can be done with only a tiny amount of oil.

The gas chromatographic analysis of the mixture of FAME takes advantage of the fact that each FAME has a different boiling point, which allows them to be separated from one another as they travel through the separation column. Shorter FAME with lower boiling points travel more rapidly through the column and will come off the column before the longer ones that have higher boiling points. Within those groupings, FAME with more double bonds (indicating a higher level of unsaturation) will come off earlier than those with fewer, or none. Once separated, the mixture can be compared to known compounds for identity or analyzed by mass spectrometry.

One especially useful quality of gas chromatography is that the response of the detector that sits at the exit port of the column is proportional to how much of each of the components is present. Thus, a larger peak will correspond to more of one component and a smaller peak to less. Often, chemists will add a known amount of an extra compound to the oil to act as a standard with which to compare the response of the natural components, giving a very reliable number for quantity.

The amount of fatty acid mixture needed for this kind of analysis is extremely small. Even a tiny bit of residue left in a pot excavated from an archeological site can give results.

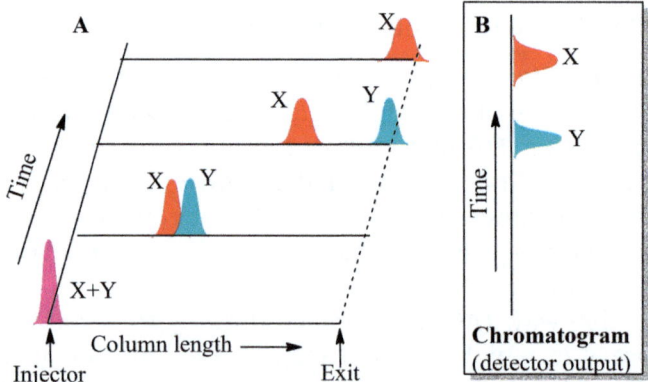

Figure 6.5 Evolution of a gas chromatography (GC) separation. GC works by injecting a sample at one end of a narrow glass column coated with high-boiling liquid, then passing inert gas through the column. Part A shows four snapshots of the GC separation: at injection of the mixed sample (bottom), as the components are partly separated, then fully separated (compound Y is exiting the column), then as compound X leaves the column (top). In part B, the results of this separation as seen by the operator: as the components exit the column, they are detected and produce a signal that depends on the amount of material present. Since compound X is retained longer than compound Y, its peak appears at a later time.

6.5.3 Best Tests for Identity: Steroid Analysis

Steroids in plants are different from those in animals. Olive oil contains about 0.2% steroids, with the largest contributor being β-sitosterol, a common plant sterol found in a wide variety of plants.[6] Of the remaining portfolio of steroids, three deserve further mention: erythrodiol, brassicasterol, and campesterol (see Figure 6.6). The latter two are named after their most convenient natural source, canola oil (*Brassica campestris*), so you might get a clue about why we want to look for them in our olive oil.

The presence of erythrodiol, which is found more in the skin of the olive than in the flesh, is a useful signal for illegal processing (specifically, solvent extraction). Elevated levels of this steroid (sometimes called a diol due to the adornment of the steroid) indicate a more thorough extraction of nonpolar components than can happen naturally.[7]

Brassicasterol is not found in virgin olive oil and therefore makes a useful marker for adulteration with common oils, such as mustard or canola oil. Finding brassicasterol in olive oil at significant levels confirms a mixed origin for the oil.

Campesterol has been used in a similar fashion as a marker for adulteration, as it is also well represented in corn and canola oil. The problem in this case is that it is also found naturally in olive oil in amounts that depend on the cultivar, harvest timing, and location of the grove.[6] Setting an upper limit on campesterol is much more difficult, especially as new regions are explored for production. Limits on campesterol are therefore under review by the IOC and already have been adjusted in other countries due to the variation in natural occurrence.[8]

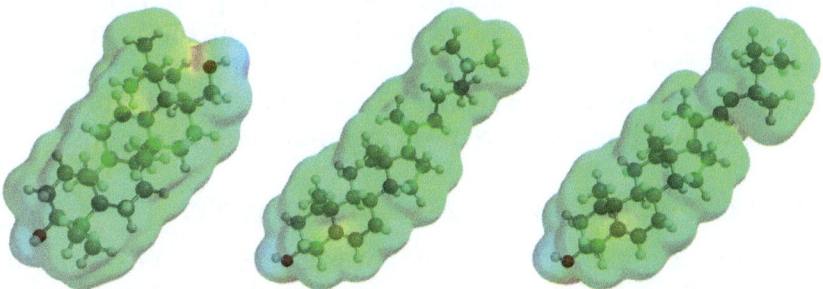

Figure 6.6 Common steroids naturally found in olive oil. Left to right, erythrodiol, brassicasterol, and campesterol.

We should mention that the steroids for which limits are given are not dangerous and that many are considered to be beneficial.[9] Still, extra virgin olive oil should not contain any chemicals – even if they are beneficial – that are not present in the pressed fruit.

Analyzing for steroids ultimately uses the same technique as for the fatty acid profile.[10] However, because the concentration of steroids in olive oil is so low, the bulk of the oil must be separated and removed before analysis. This is done by breaking down the triglycerides so that they become water-soluble, using essentially the same reaction used to make soap, as will be explained in Chapter 9. After the soapy stuff is washed away, the rest is separated using a solid preparative chromatography technique to isolate the mixture of steroids from the rest of what's left.

In addition, the steroids (solid in their pure form) have a bit too high a boiling point to be good candidates for gas chromatography. However, a different chemical reaction can add a special chemical group to the steroid to reduce its boiling point. After that modification, steroids can be analyzed by gas chromatography. Again, the behavior of compounds is tested against standard molecules or is measured in the mass spectrometer.

6.5.3.1 Steroids, Sterols, Diols, How Do They Differ?.   When discussing steroids, several terms are used, depending on the focus of the argument. We have chosen the more general term steroid, which refers to the general way the carbons in the molecule are arranged (they all have approximately 30 carbons, arranged in four fused rings with a group of about seven to nine carbons usually presented as a loose "tail" but occasionally stitched back into a fifth ring). The term "sterol" refers to compounds that have the characteristics of steroids, but have one alcohol functional group. In this way, they are the same as cholesterol, which has "sterol" at the end of its name. Finally, diols have the basic steroid structure but also have two alcohol functional groups.

6.6   HOW DO CHEMISTS KNOW THE QUALITY?

As mentioned above, the two main challenges to ensuring the quality of the oil you buy are reading and understanding the label and determining if deterioration has occurred after bottling but before purchase. Because they have much the same effect, we'll treat them together.

Another tool for analyzing olive oil is sensory analysis. An oil that is of poor quality or has decayed due to time or bad treatment will likely have a sensory defect, such as "rancid." We'll discuss this type of analysis thoroughly in the next chapter.

Cameo 6.1   Jamie Ayton, Oil Chemist, Wagga Wagga, Australia

Jamie Ayton is a chemist at the NSW Department of Primary Industries Australian Oils Research Laboratory, in Wagga Wagga, Australia. This state of the art facility, charged with assuring the quality and safety of all edible oils, is run with precision and integrity. The globally recognized center has been accredited through the International Olive Oil Council since 2001. It has clearly defined protocols, well-trained personnel, and a track record for reliability and reproducibility. The instrumentation we saw during our May 2015 visit is impressive. Jamie told us that the lab can do more than 20 different chemical tests on olive oil and has longstanding projects on the effects of storage, irrigation, and timing of harvest on the quality of the oil. His broad knowledge of oils in general was shared with us and helped us place olive oil in the context of other agriculturally derived food oils.

We enjoyed our visit to the legacy Colonial trees on the campus of Charles Sturt University at Wagga Wagga. In South Australia there are many century old olive trees, known as Colonials, still bearing high-quality fruit. In fact, the history of the olive tree in South Australia goes back 165 years to when George Stephenson, the secretary to Colonel Light, imported several trees from the Mediterranean region. Since olives prefer temperate climates, the areas around Adelaide and South East Australia have ideal climates for olive production.

6.6.1   Best Tests for Quality: Free Fatty Acids (FFAs)

A free fatty acid (FFA) test, often called "free acidity," is among the older tests done to establish the quality of virgin olive oils. It is simple and can be done in a small lab space associated with the processing plant or storage facility. It is a simple titration, in which a measured amount of oil is mixed with solvent and an indicator, then increasing amounts of a reagent (a common base such as sodium hydroxide) are added until the indicator changes color, showing that the amount of base added matches the amount of FFA in the sample. One can then calculate how much of the sample has deteriorated to produce these FFAs.

FFAs are produced by the reaction of TAGs with water. Normally, this reaction is very slow, but if accelerated by enzymes, it can produce measurable amounts of FFAs. The enzymes are separate from the stored oil in an intact olive, so with careful processing at a lower temperature, the resulting oil has very low FFA content (often 0.2–0.4%). However, any damage to the fruit can cause the oil and enzymes to mix. This damage can include physical damage such as crushing or the hole created by the olive fly, or microbiological damage such as mold or bacteria. Olives held at higher temperature will at least soften, if not burst, and will likely give a higher FFA content. Similarly, the longer the olive sits in its damaged state, the more this reaction can occur. Olives sitting on the ground for a week will produce oil with extremely high FFA levels (we have seen containers marked 12%, but assume it could be higher). Such oils are not fit for consumption, but can easily be used to make soap.

In the same way that a high peroxide value is a marker of a troubled oil, but does not itself affect the taste, the FFAs themselves do not have an identifiable flavor. They do, however, indicate damage to the olive structure and are a useful and easily determined indication of quality.

6.6.2   Best Tests for Quality: Diacylglycerol (DAG) Analysis for the Age of the Oil

The breakdown of a triglyceride produces two products: one is the FFA and the second is the parent molecule minus the FFA. We call the remaining part a diacylglycerol, or DAG for short. The structure of a DAG is shown in Figure 6.7.

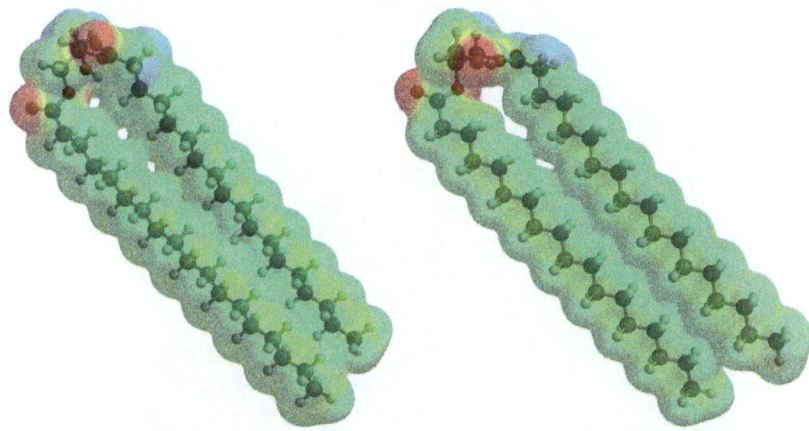

Figure 6.7 Diacylglycerol (DAG) is composed of two fatty acids connected to a glycerol molecule and can be put together in two ways. (Left) The two fatty acids are connected at positions 1 and 2 of the glycerol (1,2-DAG) and (right) the two fatty acids are connected at positions 1 and 3 of the glycerol (1,3-DAG). Note the extra space between the chains in this molecule.

Generally, when DAG molecules are first made from TAGs, the outermost fatty acid has broken off and the remaining two fatty acid chains are right next to each other, bound to oxygen atoms at positions 1 and 2 in the glycerol molecule. This molecule is called 1,2-DAG, as shown in Figure 6.7(a). The two chains are a bit crowded, so over time the fatty acid in the middle can migrate to the other end of the glycerol to become bound to the oxygen at position 3, making a new molecule called 1,3-DAG, as shown in Figure 6.7b. This reaction does not depend on other molecules to run, thus giving us a clock that depends only on temperature to run at a known rate. We can measure the amount of 1,2-DAG and the amount of 1,3-DAG by liquid chromatography and determine the ratio of 1,2-DAG to the total DAG (1,2-DAG/[1,2-DAG + 1,3-DAG]). The higher the ratio, the fresher the oil.

6.6.3 Best Tests for Quality: Pyropheophytin (PPP) Analysis

Often, oil processed from green olives will have an emerald green color when first pressed, as shown in Figure 6.8. This color will fade quickly as the oil ages. When you look at a chemical sample for its age, it is often true that there is some component in the

Figure 6.8   Freshly pressed emerald green olive oil from Picual olives from Rangihoua Estates (Waiheke Island, New Zealand).

mix that will decay at a predictable rate. Several years ago, scientists studying the pigments in olive oil came up with the idea of looking at the breakdown of chlorophyll and its relatives.[11,12] Pigments are very nice to study because they are easy to detect (by definition they absorb light in a characteristic way). This is also consistent with what fans of olive oil have known for some time: if you leave an oil out for some time, the color will change.

To make this system work, scientists had to understand the breakdown pathways of chlorophyll, which match the pathways in many other green plants. As described in Chapter 5, chlorophyll in the oil is actually short-lived, and loses an atom of magnesium from the center of its ring to produce pheophytin (PP). Most of the pigment in a green oil is PP, which will gradually decompose into pyropheophytin (PPP). The structures of both these pigments are shown in Figure 6.9.

In a fresh oil, there is very little PPP, which is more yellow than green. The conversion of PP to PPP is a simple reaction that does not require any other chemicals to occur, only sufficient energy. Therefore, the rate of the reaction, and thus the amount of PPP, is governed only by the time since production, and the storage temperature and light exposure (higher temperatures and stronger light accelerate the reaction). As time goes on, the amount of

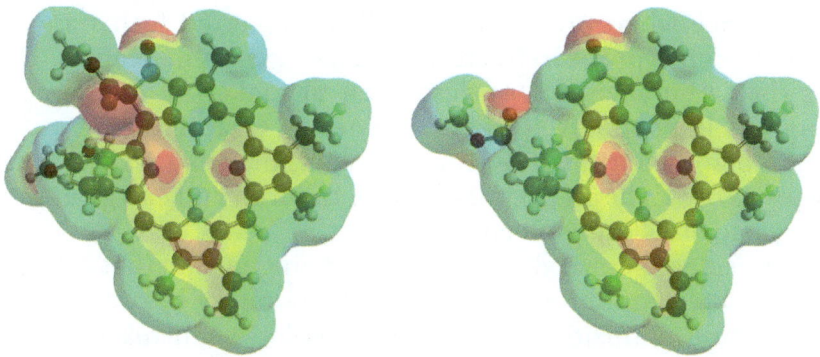

Figure 6.9 Chlorophyll breakdown products in olive oil, pheophytin PP (left) and pyropheophytin PPP (right). Each has a long hydrophobic tail, hidden behind the face of the molecule. Note that the section of PP in the upper-left corner is missing in PPP.

PP will decline, and the amount of PPP will increase. Chemists can test this relationship by separating the pigments using chromatography and finding out how much of each pigment is present. Since each oil starts with a different amount of PP, the result is corrected by calculating the ratio of PPP to total pheophytins (PPP/[PP + PPP]). In general, the lower the ratio, the fresher the oil. As the oil ages or is treated poorly, this ratio will rise. At some point, all of the PP will be destroyed (and perhaps all of the PPP), making the ratio meaningless. That's really the worst possible outcome and indicates a very poor quality oil.

Both the PPP and the DAG tests are fairly linear with time, and so make a reasonably good clock for oil age. In contrast, the next two tests say more about the qualities of the oil, without giving the exact age. A younger oil, poorly treated, may be of worse quality than an older oil kept in better storage conditions.

6.6.4 Best Tests for Quality: Peroxides Test for Oxidation

When we talked about antioxidants, we described the kinds of damage that can be done to triglycerides by molecular oxygen. Some of these products can build up in an oil that has been stored in the presence of oxygen. In particular, chemists are interested in molecules called organic peroxides (see Figure 6.2 for an example). These are typically fatty acids that have had molecular oxygen added at some point along the chain.

The place where the oxygen has been inserted forms a weak spot which is prone to fracture to make smaller molecules. This is the same chemical process used to make the aroma compounds in good olive oil, as explained in Chapter 5, but rather than being controlled by enzymes (that make sure the oxygen goes to the right place), this is more random and can give some very bad smelling molecules.

Peroxides themselves are good oxidizing agents that can easily be detected in the quality control laboratory by presenting a fast-acting antioxidant that changes color when reacting with the peroxide. The more intense the color, the more peroxides are present. These test chemicals are usually calibrated with a known peroxide, letting us measure the amount of peroxide present by comparison. Results from this test are presented on some labels as "PV" (peroxide value) and are usually reported as an equivalent mass of hydrogen peroxide. The upper limit of the peroxide value according to IOC regulation is 20 units per kg. Without some reference, this number can be difficult to evaluate, but generally a larger number is worse than a smaller one. The actual level of peroxides in oil does not simply increase as exposure to oxygen increases, but depends on a cyclical wave of peroxide formation and decomposition that develops in the stored oil.

6.6.5   Best Tests for Quality: UV Absorption Tests for Rancidity

Properly stored olive oil takes a long time to achieve noticeable rancidity. We have heard stories of oils that have been relatively free of rancidity after even five years of controlled storage. However, this story came with the clear warning that the oil was not what it once had been: the positive attributes like grassy and apple aromas had seriously deteriorated. In addition, this oil had been stored very carefully at lower temperature in the dark. The usual bottle in the supermarket has not been stored that carefully. A clear sign of deterioration because of age is an oil that smells like stale nuts or crayons.

One of the consequences of attack on fatty acids by peroxides is a modification of the unsaturations (double bonds) in the chain. This can take one or both of the following forms: migration of an unsaturation (double bond) from one position to another or change in the arrangement of the atoms around the double bond from *cis* (non hydrogen atoms attached on the same side of the double bond) to *trans* (non hydrogen atoms attached on opposite sides of the double bond). This is especially pronounced in PUFAs, as they are both more susceptible to attack by peroxides and have more unsaturations (double bonds) to work with. This activity can take place whether the fatty acid is free or in a triglyceride.

Let's look a little closer to see what is happening. Figure 6.10 shows a fragment of the doubly unsaturated linoleic acid that we'll use for this discussion. Ordinarily, the unsaturations in fatty acids are separated by one carbon atom which is not involved in either double bond. In natural linoleic acid, we see an unsaturation between carbon atoms at positions 9 and 10, and a spacer carbon atom at position 11 that is not involved in any unsaturation, and then an unsaturation between carbon atoms at positions 12 and 13. Both unsaturations are also *cis*. Attack by the peroxide can remove one atom of hydrogen from the carbon at the 11 position, leaving that carbon unstable with only three bonds. The electrons in the double bond between 9–10 and 12–13 respond to this by delocalizing – or spreading out over the five carbon atoms to give the unstable form shown in the center of Figure 6.10. If that molecule is lucky enough to find a hydrogen atom to replace the one lost (it may find it in another molecule of linoleic acid), it is most likely to replace it on the end carbon, either 9 or 13. This gives a new arrangement in which the unsaturations are not separated by a spacer atom and the arrangement has changed from *cis* to *trans*. Notice that the double bonds are conjugated now, just as they were in the peroxides discussed in Figure 6.2.

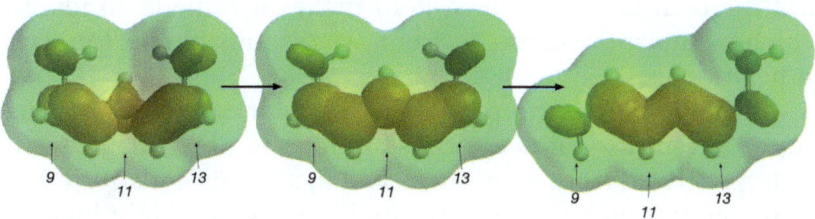

Figure 6.10 Rearrangement of the double bonds in linoleic acid as a measure of rancidity.

While the original fatty acid does not absorb ultraviolet light (UV light), the new form in which the bonds are connected does. The absorbance is deep in the ultraviolet range (at about 232 nm), but well within the capacity of modern instruments. A similar reaction that takes place with linolenic acid (three unsaturations) can give rise to a more extended conjugated structure that absorbs at about 270 nm. The absorbance of oil at 232 nm and 270 nm is extremely sensitive, easily measured, and has now become a standard method of quality control for oil. Referred to as K232 and K270, these values can tell us about how much oxidation has taken place.

Both of the standard values (K232 and K270) have numerical limits attached to them, and a good oil should test below that limit. To help give a more concrete meaning to each value, we have estimated that these upper limits correspond to approximately 1 damaged molecule in 25 total (linoleic acid from K232 or linolenic acid from K270).

This type of testing generally comes with a third measurement, called the ΔK value. This is not a comparison of K232 and K270, but a separate measure of the area around 270. Essentially a test for the shape of the curve in that region, it is sometimes used to uncover the presence of refined oil. In authentic oil, it involves a comparison of very small values, so it is not considered reliable as a measure of the quality of extra virgin olive oil (EVOO).

6.6.6   Another Test for Quality: Antioxidant Content

We strongly believe that antioxidant content is an important part of describing olive oil. Such a key contributor to both the flavor and the health effects of olive oil deserves careful measurement. There is no universal correct value for the antioxidant quantity and the complexity of the system makes it difficult to identify which of the antioxidants present are responsible for the flavor or health effects.

As mentioned in the processing chapter, some of the most important antioxidants are those unique to olive oil. The compounds derived from oleuropein and ligstroside constitute a big family, containing dozens of compounds. As you might imagine, there are two ways to quantify these compounds. We could simply ask what the total antioxidant capacity is, without trying to figure out the function of each. That's a bit like measuring how loud your family reunion is. A loud family reunion could be either lots of quiet relatives or a few loud ones. On the other hand, one could find out how much of each antioxidant is present, then add up the individual contributions. The first test is less specific, but faster, and the second type is slower, but will give a better idea about antioxidant contributions in the long run.

To measure bulk antioxidant capacity, the oil is treated with an oxidant that changes color when it reacts with the antioxidant. The chemicals are chosen to give a linear color change with this reaction, and the results can be read by a spectrometer. The results of this test are usually compared to hydroxytyrosol, and are given as mg of antioxidant per kg of oil. It is worth noting that the oxidant that is chosen is not found in humans (not healthy ones, anyway) or in the air, so the relevance to health or shelf life is more speculative. It may be performed quickly, however, and can give a good, crude measure of antioxidant content. There is a commercial kit being sold that allows producers to make this measurement in their facility.

Because of the common structural features of the olive oil antioxidants, they can be extracted and separately quantified with relative ease. Using a short column which captures the antioxidants from oil passed through it, but releases them when a different solvent is passed through, the compounds can be purified and concentrated for analysis by high performance liquid chromatography (HPLC), which separates the compounds for individual quantification.[13] This gives a profile of the oil and lets us know how much of the antioxidant power is due to the more powerful oleuropein relatives, or the less powerful ligstroside relatives. It should also allow the measurement of oleocanthal, responsible for many health effects detailed in Chapter 8. This is a much more time-consuming and expensive process, but it gives a better indication of quality.

The antioxidant profile is sufficiently important to inspire the development of other techniques as well. Seeking a technique requiring little preparation of the sample, several researchers have proposed the use of nuclear magnetic resonance (NMR), which uses very high magnetic fields and radio waves to probe the chemical environment of atoms in molecules. Many of the antioxidants can be measured simply by analysis of a spectrum taken in a few minutes with this instrument.[14–16] The expense of the instrument may hamper adoption in the industry, but it is a very powerful technique.

6.6.7 Feel Free to Put Your Oil in the Refrigerator – But It Won't Tell You the Quality

There is a common belief that putting olive oil into the refrigerator will help determine the quality. Supposedly, if the oil solidifies, it's good, and if it doesn't, it's bad. Chemists who run the time-consuming and expensive tests for quality wish this were true, as it would save a lot of time and effort.

Unfortunately, the people who came up with the theory weren't chemists. If they were, they would have known that a mixture of compounds (as olive oil is) always freezes at a lower temperature than a pure compound. Pure triolein, the triglyceride made from three oleic acid molecules, freezes at 5 °C, or about 40 °F. This is slightly above the temperature inside your refrigerator. The waxes present freeze at even higher temperatures. Since real olive oil is a mixture, it will not freeze solid unless placed in the freezer. You might see some partial freezing, and the bottle may look slushy, but it will likely not freeze completely.

At the end of this experiment, should you try it, you can get your desired oil back by just letting it warm to room temperature. No harm is done to the oil. You will find that it takes quite a while to thaw the oil, which explains why people don't usually store oil in very cold conditions.

6.7 HOW DO CHEMISTS KNOW WHERE IT'S FROM?

Describing the origin of the olive oil in a bottle can be a bit more complicated than it might appear. It's simple if you drive to a grove, see the trees that have been harvested, look at the press on the property, and buy the oil directly from the producer. In that case, the origin of the olives, the pressing, and the bottling are all on one site, and only one location need be mentioned.

On the other hand, it's possible for olives to have been grown in one location, pressed in another, and bottled in a third. Adding to the complexity, the bottlers may blend oil from several different locations (even different countries). How can we make sense of this?

If the oil is indeed extra virgin, one factor in our favor is that it is nearly impossible to ship olives long distances to a far distant press without losing too much quality. Good quality oil requires that the grove and the press be in the same region.

The same can't be said for bottling. With good storage conditions (blanketed with nitrogen in a stainless steel tank at a reasonably low temperature), the oil can be shipped easily to a destination days away. In discussing origin, we should clearly distinguish production from bottling.

When it comes to an oil that has been bottled in one country from oils blended from several other countries, most countries require this information to be disclosed on the label.

6.7.1 Does it Matter?

We should be clear that there is nothing inherently wrong with olive oil blended from many different regions, as long as it's done well. Especially if the countries of origin are listed on the label, a good-tasting oil from multiple origins suits the needs of most consumers. It can be hard to find this information, however. Looking on the label and finding the phrase "Bottled in Nostrovia" will not necessarily tell you where the oil comes from. Even "Product of Nostrovia" might be allowed for oil bottled in that country.

Within the past few years, labeling requirements in some of the biggest import markets (USA and EU) have required that the origin of all constituent oils be indicated on the bottle. Properly labeled oil should say something like "Imported from Nostrovia, containing extra virgin olive oil from Europia, Lunaria, and Saturnia." (Please note – Nostrovia is not a real country, but a misspelling of the Russian "Na Zdorovie," meaning "To your health!" Europia, Lunaria, and Saturnia are similarly fictional countries created for this example.)

6.7.2 What are the Best Tests to Apply from a Chemist's Point of View?

Determining the country of origin for a particular olive oil or countries for a blend of oils can be quite a challenge. There is no single test that can establish where an oil is produced.

However, it is possible to run an array of tests and use some good detective skills to establish the origin of an oil. This kind of work is typically reserved for cases of fraudulent labeling, as it is difficult, expensive, and time-consuming.

6.7.3 Best Tests for Origin: Multicomponent Analysis

One trick up the chemists' sleeves is a statistical analysis of data often referred to as multicomponent analysis. It requires considerable background work before the analysis can be performed, but once this is done, a quick analysis is possible. Chemists might take a number of readings from a spectrometer (for example a machine looking at the way infrared light interacts with the olive oil) or from the analysis of the contents (like the fatty acid or polyphenolic profile). Using oils of known origin, these readings can be combined into a few scores, which can be used to locate the origin of the oil. The chemists designing these tests look for scores which reliably group oils of the same origin together and oils from another origin together, but with different scores.

6.7.4 Best Tests for Origin: DNA Tests

Dedicated viewers of police procedural shows will know that one can find DNA in almost any sample. The same is true for olive oil. Even though DNA is not very soluble in oil, there are nevertheless small amounts present, and even a few molecules of DNA can be amplified and analyzed. That should provide proof positive of the cultivar, which is very helpful for locking down the origin of the oil. Also, the presence of, for example, soybean or sunflower DNA in the sample is clear evidence of an oil which is not pure olive oil.

This practice has been greatly aided by recent developments in the extraction of DNA and new studies that have identified which part of the DNA is most helpful.[17] Another issue is that the genetic information even for a single cultivar might be a mixture of two sources. Since some cultivars do not self-pollinate, the oil will contain genetic information from the pulp of the olive (provided by the parent tree) and from the pit (provided by the pollinator).[18]

In fact, research in DNA analysis of olive oil is on-going, as shown by the recent publication of the complete genome of the olive tree, discussed in Chapter 2.[19]

6.7.5   What Practices Affect Authenticity?

Authenticity can be significantly affected by producers' honesty and book-keeping procedures. Good producers keep careful records of when oils were produced and from which groves, of amounts blended to make a batch, and how many bottles were produced from that batch, and they need to make sure the numbers are consistent. This is not glamorous work, but it is crucial to maintaining the reputation of the producer. Someone who takes 5000 liters of oil, blends it to make 10 000 liters of product and calls it "extra virgin olive oil" will not have an authentic product.

Of course, storage and transportation of product is crucial to maintaining quality. This includes not just the producer and shipper, but also the distributer and retailers. This supply chain is only as strong as the weakest link. Fortunately, the methods for proper shipping and storage are not especially difficult.

6.7.6   National and International Standards

In international trade, clear and specific standards must be developed that are informed by long experience with what makes good olive oil. Usually, minimum standards are set for various grades of olive oil. By stating minimum standards, and rigorously establishing measurement techniques for measuring them, international standards can be used in legal battles such as those that might ensue when a country is faced with the import of improperly labeled oil.

6.7.6.1   What are the International Olive Council (IOC) Standards and How Do We Test for Them?.   Each country or trade group will have its own standards. This is certainly true for the US, the European Union, and many other countries.[20] However, most of these are informed by and rely upon standards propagated by the IOC, which keeps and maintains both standards and standard methods for analysis. This is an intergovernmental organization representing countries providing 98% of the world's supply of olive oil, in addition to representing table olive producers. Member countries are widely distributed and include 28 countries in the EU and 14 countries outside it. Another source of standards and definitions comes from the Codex Alimentarius,

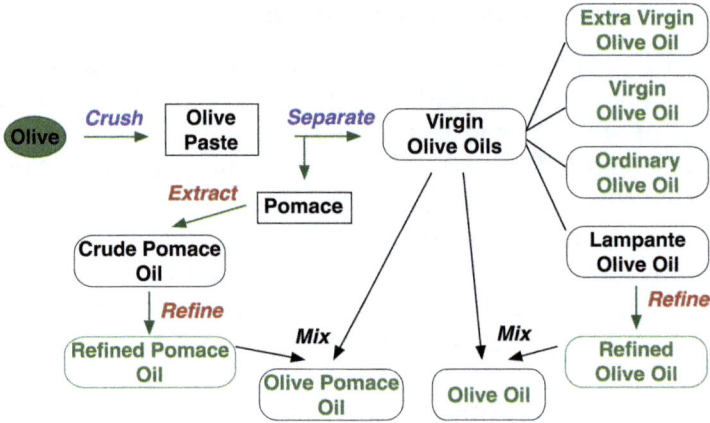

Figure 6.11   Grades of olive oil. Edible grades are in green, while inedible grades are in black. Note the near duplicate appearance of "virgin olive oils," which is a general category describing all oils obtained by mechanical means only, and "virgin olive oil," an individual grade within that category.

covering all foods, produced by the Food and Agriculture Organization (FAO) and the World Health Organization (WHO).[21] The Codex standards are today considered the authoritative standard by the World Trade Organization (WTO).[22]

Nine grades of olive oil are recognized by the IOC, as illustrated in Figure 6.11.[23] Four are made from olive fruit by purely mechanical means and five are made from olive products (fruit, oil, olive waste, or pomace) by chemical modification or extraction processes. Each of these grades is described by a strict set of criteria that includes chemical analysis and organoleptic (tasting) tests set out by the IOC as well as national and local agencies. In the USA, the national agency maintaining standards is the Food and Drug Administration (FDA).[24] State laws in the USA will sometimes add additional criteria that govern labeling and analysis. The four grades of oil that come from purely mechanical pressing or centrifugal spinning of crushed olive fruit (often confusingly referred to as "virgin olive oils") are *extra virgin, virgin, ordinary*, and *lampante* oil. The last of these, lampante, is not considered fit for human consumption. Three grades of oils derived from chemical treatment or refining of the olive fruit, oil, or waste products (olive pomace) are classified by the IOC as *refined olive oil, crude pomace oil*, and *refined pomace oil*. Two mixtures derived

Table 6.1 International Olive Council (IOC) standards for grades of virgin olive oil.

Grade	Free fatty acids	Peroxide value	K-232	K-270	Sensory positive	Sensory negative
Extra virgin	≤0.8%	≤20	≤2.50	≤0.22	>0	0
Virgin	≤2.0%	≤20	≤2.60	≤0.25	>0	<2.5
Ordinary	≤3.3%	≤20		≤0.26		≥2.5, <6
Lampante	>3.3%	No limit				≥6

from the classes above are also available: *olive oil* (a mixture of refined and virgin oils) and *olive pomace oil* (a mixture of olive pomace and virgin oils).

As mentioned throughout, the highest quality of olive oil is "extra virgin." There are other terms in other countries (for example, an oil that is extra virgin oil is labeled "Naturel Sizma" in Turkey), but this product is generally similar worldwide. Oil that is from a process similar to the extra virgin, but which doesn't meet all the standards is called "Virgin" olive oil in the USA. This oil can have many of the positive attributes of extra virgin oil, but generally does not taste quite as good.

These standards include both chemical and sensory evaluation of the oils. Sensory analysis (described more fully in Chapter 7) evaluates both good sensations and bad, also called positive or negative attributes. In short, extra virgin olive oil must have a minimum score for positive attributes and a zero score for negative ones, in addition to the lowest amounts of impurities in chemical testing. Table 6.1 gives the IOC limits for the four grades of virgin olive oil.[21]

In addition to these basic standards, limits for a variety of other characteristics are provided by the IOC. The types and amounts of steroids (called sterols), the composition of fatty acids present, the amount of *trans*-fatty acids, and many other characteristics are specified.

6.8   CERTIFYING QUALITY: TESTS FOR OLIVE OIL

Once the oil is packaged and ready to ship to its final destination, several parties have a strong stake in the outcome. The producer, the shipper, and the wholesaler will be interested in maintaining a reputation for being able to provide high-quality and safe products, and the consumer will want an authentic, delicious, and stable oil.

Who exactly is in a position to ensure a positive outcome for all? Who does the testing? There are two types of laboratories: ordinary contract laboratories and accredited laboratories. Both can ensure high-quality results. However, only accredited labs are checked by an outside agency. To maintain accreditation, a lab must test a blind sample sent from the outside and match accepted values. This must be done for every technique being certified. It's also not enough to do this once, but the certification must be renewed on a regular basis. Some certifications are for practices you might not even think about, such as the tracking of samples and data storage practices. It's a lot of work maintaining certification, and the certified labs we visited were justifiably proud of their achievement.

In speaking with managers of testing laboratories, we frequently heard the story of producers, new to the business, who sent samples of their oil asking for "all the tests." When informed that the complete panel of tests would run several thousand dollars, there might be a long silence on the telephone until the manager told them that for casual sales at a farmer's market, or sales within the country of origin, the testing could likely be done for just a few hundred dollars or less. In truth, the list of available tests is very long, and what is required depends on where the oil is going. Selling to your friends requires just the basics, if that. Exporting to China requires almost the entire list. Fortunately, the tests can be done on a large batch of oil, before bottling, so the expensive testing is reserved for large batches being exported.

6.9   WHAT THE LABEL TELLS YOU (AND DOESN'T)

6.9.1   Claims of Quality

If you want the highest quality olive oil, it must say "extra virgin" (or the local equivalent) on the label. Anything else will not deliver on its promise of good flavor, absence of defects, and good lab tests. Among extra virgin olive oils there is a wide variety of qualities which may not be captured by the name, because "extra virgin" is really a minimum standard. It is still important to learn a bit more about the oil, perhaps by talking to the seller or by tasting the oil yourself.

There are two types of olive oil that are distinctly different from extra virgin and virgin. Both are the result of a more intrusive process and lack almost all the unique chemical qualities of extra virgin oil. The first is "refined" oil, and the second is "pomace" oil.

"Refined" oil certainly sounds like it should be superior in quality to other oils. In fact, as you might recall from the discussion in the previous chapter, it is chemically altered: the antioxidants and free fatty acids are stripped out, the oil is deodorized and decolorized, and we are left with a nearly colorless, odorless oil. Producers often add in some naturally produced olive oil (most likely virgin and not extra virgin) to give some flavor and color. On labels, this is sometimes referred to as just "olive oil" (look for the absence of "extra" or "virgin"), or even "pure olive oil," a misnomer made possible by the fact that it has fewer components than extra virgin oil. In the USA, it is often labeled as "extra light olive oil."

Indeed, there was a time when the quality of oils produced was very low and export markets not very discerning. Rather than sell an oil that was high in acidity and had taste and aroma defects, the chemical industry developed a process called "refining" in which the impurities were removed. Similar "purification" takes place in the processing of other foods such as flour.

Refined olive oil can be useful, for example, if a particular country's cuisine is not compatible with the flavor of olive oil or if an individual doesn't like the taste. It is also possible to cook foods at a higher temperature with most refined oils.

One category of particularly low-grade olive oil is called crude pomace oil. This is extracted from the olive waste by the use of solvents such as hexane, which dissolve the oil and allow an easy separation from the waste. After the extraction, most of the solvent is removed. Crude pomace oil can also be refined, or used as-is for technical uses such as lighting or soap.

6.9.2 "Best by..." Date

Most carefully labeled olive oils now come with a "Best By" or "Best Before" date printed on the bottles. These dates are calculated in several ways, but all are useful to the consumer. It makes sense to seek the bottles with a "Best By" date as far in the future as possible.

The dates on the bottle are only one aspect of your decision, but they can be helpful in weeding out confusing choices. If you examine a bottle of olive oil carefully, you might find a text such as "Best by Nov. 22, 2022," or "Best Before 11/11/2018." As is the case with many foods, it may well be safe to consume the product after the date printed on the container (or may not be), but the date is a strong indication that, after the date, the product will not meet the quality and flavor standards promised by the label.

The "Best By" date is calculated by the bottler, sometimes by adding two years to the bottling date, regardless of the stability of the oil at that time. Perhaps this accounts for some of the failure of some to meet the extra virgin standard by bottles of oil found on retailers' shelves. Other parties may be more conservative and calculate the date from the time of harvest. A simple calculation based on geographical location will provide some indication of this. An olive oil from the northern hemisphere oil bottled in July will not have been produced in July, which is already in the next year's growing season. Most likely, it was harvested the previous year. A southern hemisphere oil bottled in July might be at the very beginning of the bottling for the season, and will be quite a bit fresher.

Some producers use accelerated aging techniques to predict the stability of the oil. If done carefully, in an experienced lab, this can ensure confidence in prediction of the relevant date. In the end, the best use of this date is to ensure it is as far away as possible. Fresh oil is always better than older oil.

6.9.3   Certifications

Often a label will carry a certification description, or a separate sticker will be applied to the bottle. This means that the oil meets the standards of that organization, whether it is for the entire country (South Africa, Australia, for example) or for a more local region (California, for example, has its own standards). Generally, lab tests must be done by a lab recognized by the organization and sensory analysis performed by a recognized and usually certified tasting panel.

6.9.4   Origin, Cultivar, and Style

In many cases, the country or region of origin contains the whole story of the oil: growing, production, storage, and bottling. In some cases, especially in the Mediterranean region, oils from different countries are blended to achieve an oil meeting standards for flavor and chemistry. These oils could be of different ages as well, as oil from a year with excess production may be stored to offset shortages the next year, for example. As a result of increasing consumer pressure, much clearer labeling of the country of origin is being required in a number of countries.

The disclosure of varietal is optional, and generally reflects a marketing decision. The relationship between the variety of olive and the characteristics of the oil it produces is difficult for many consumers to discern. One way of evaluating this description is the indication of the "intensity" of the oil, using words such as "mild," "medium," and "robust." These often relate to the bitterness and pungency of the oil, and are based on sensory notes from a tasting panel or trained taster.

6.9.5 Cold Press?

Even in the age of high-tech centrifuges for the separation of oil from the pomace, we still see the words "first cold press" on labels. This is a nice nod to the history of oil production, even if not technically accurate. As you recall from Chapter 5, in a modern mill there is only one separation (so is technically the first, but there is no second extraction). The temperatures are not what we would describe as cold, but rather controlled to remain below 27 °C for the entire production process, which allows the production of excellent oil with a decent yield. The older olive presses have given way to continuous flow centrifuges, which are more efficient and give higher quality oil. Other terms used include "cold extracted" or "cold pressed" and similarly refer to the industry standard of keeping the process temperature relatively low. These phrases are a reminder of the extremely long connection between olives and humans.

6.9.6 Lab Results

In many countries, indication of lab results on the label is optional, or only a small set of results is required. Now that you know what these results mean, you understand that few of these results are going to tell you what the oil tastes like. However, some may tell you a bit about how the oil may age. Olive oils with higher phenolic content are typically more durable than those with low content, all other things equal. As we will see, they are also usually more bitter and pungent, depending on the profile of compounds.

6.9.7 Wide Variety of Labeling Styles

As any company that exports a product knows, the labeling requirements are specific to the country (or region in the case of the European Union which has established standards relevant to all (currently) 28 member states) in which the product is sold. In addition to language requirements, the information which must be disclosed may differ substantially.

Country of origin: Fortunately in the USA, the label on every olive oil container must contain information about where the oil came from, or the "Country of Origin Declaration." This is enforceable by the FDA or, if the oil is imported, by the US Customs. The US Tariff Act of 1930 as amended (19 USC s. 1304) requires that

> "...every article of foreign origin (or its container) imported into the United States shall be marked in a conspicuous place as legibly, indelibly, and permanently as the nature of the article (or container) will permit, with the English name of the country of origin."

6.10 THE ROLE OF THE RETAILER

Retailers are becoming more conscious of the specialized characteristics of olive oil and the special needs for storage and display. That said, there is still progress to be made in the storage and display of olive oil. Cases of olive oil in a hot storeroom, or clear glass bottles on a top shelf (near the bright lights of the store) will decay more quickly than if more carefully stored and displayed. Buyers can be faced with a bewildering array of choices among different oils, with little time to make the choice amongst the myriad of other buying decisions they must make.

However, like any other product, wholesalers and retailers are best advised to choose oils based on their flavor, to store them properly (ideally below 20 °C and dark) and to test oils periodically if they have been stored for many months. We hope this chapter is also useful in expanding knowledge of the types of testing available to ensure that the oil is labeled correctly. As customers become more aware of what a good olive oil tastes like, good retailers will surely want to be able to meet consumer demands.

6.11 AT THE FINAL DESTINATION

6.11.1 The Role of the Restauranteur

Running a restaurant is a tough business which, if done at the highest level, requires attention to a myriad of details, among which involves the olive oil used in cooking and on the table. Many great restaurants know this, but many restaurants could improve how they handle the oil. Restaurant staff should know what a good olive oil tastes like and how to evaluate the quality. A recent study showed that 60% of the olive oil labeled as "extra virgin" in restaurant supply channels failed the sensory evaluation test for the classification and also typically failed the more sophisticated tests such as the DAG test.[25]

In addition, the heat and bright lights of a kitchen can seriously degrade oil in a short period of time. It is best to store oil in a cool and dark place, and bring out a small supply at a time. Olive oil for the table is best served fresh with each meal. If provided on the table, it should be in a container with a non-refillable top. There is a temptation to treat olive oil like salt and top up bottles that sit on the table for weeks. We hope to have convinced you that the result will be quite disappointing for all before long.

6.11.2 The Role of the Home Consumer

If you are buying the oil for home, there are some important habits to cultivate that will pay off in your enjoyment of high-quality olive oil.

In choosing what to buy, olive oil is seldom worth buying in bulk, unless you have an extremely large family. An open bottle of olive oil can decline in quality fairly quickly even if stored well. You should not plan to have more than six months of oil in an open container, and preferably much less. Good storage for olive oil means cool and dark. Right next to your cooking area is a very bad choice for anything but one day's supply. For unopened bottles, the basement might be a good choice, as it gets closer to the optimal storage temperature of about 15 °C.

Decorative containers for oil are very nice for serving oil at a single meal. Try to choose containers that are smooth and easy to clean on the inside, as they should not be used to store oil and should be cleaned after each use. It's surprising how little rancid oil it takes to contaminate a whole container.

6.11.3   Shelf Life with Proper Storage

Once you've procured your oil from a store, grove, or online and placed in a cool, dark place, how long should it last?

First, it is worth repeating that olive oil is always better when younger. Even the best oils will be diminished in quality by storage, so buy the oil to use, not to store.

There are certain chemical factors that affect shelf life. One of the most common defects in properly packaged oil is rancidity (the stale aroma that comes from the slow oxidation of oil described earlier in this chapter). Proper storage is the first line of defense, but the oil itself has some chemical defenses that allow for extended life.

First among these are the antioxidants, particularly the phenolic compounds that also give rise to the bitterness of the oil. Of these, the most potent are in the oleuropein family, as they bear two hydroxyl groups on the ring, enhancing their reactivity with oxygen.[26-28] They can also regenerate oxidized tocopherol (vitamin E) molecules, extending the life of that antioxidant system.[28] However, the antioxidants from the ligstroside family can also be effective, as can α-tocopherol. All of these compounds can protect against oxidation as long as they last. Consequently, they will decline in concentration during storage, and the bitterness will also decline in tandem.

In addition to the protective function of the antioxidants, the nature of the TAGs also plays a role. Since the saturated and monounsaturated fatty acids are less easily oxidized, an oil with more MUFA and fewer PUFA will last longer. We've seen that some extra virgin olive oils are higher in oleic acid than others, due to the cultivar and the location of grove. A high oleic acid content conveys a resistance to oxidation which, when combined with the protective effects of the antioxidants, can give a long life to olive oil. These two factors likely provide most of the protection against rancidity given to olive oil.

Oxidation of olive oil is one of the more common means of degradation found during storage, but is not the only one. For example, in unfiltered oil with sufficient moisture content, active microorganisms can be found for several months after production and can have either positive or negative effects on the flavor of the oil.[29,30] Depending on the health of the olives used to make the oil, the microorganisms might be benign, or might cause flavor defects such as "muddy sediment," "fusty," or others. While there is an accelerated method for measuring the onset of rancidity, we do not know of a similar test for the development of microbial defects, as high temperatures kill the microorganisms.

The bottom line, then, is the same as the top line: use your olive oil up and don't store it. Just one more thing...

There is one more test for olive oil quality. We have chosen to devote the entire next chapter to this test, as it is arguably the most important one. It is also wrapped up in the enjoyment of the oil and in the magic of the process followed to prepare it. We are speaking, of course, about the flavor of the oil. This flavor can be probed using laboratory tests, artificial noses, and artificial tongues, but nothing to this point beats the human senses, properly trained for evaluating oil.

After all, extra virgin olive oil should taste good. It should taste like the olives it comes from. Thankfully, this is part of the definition of an extra virgin olive oil that is true to the label. Even those who are just fans of the oil should find the steps for evaluating the oil useful.

REFERENCES

1. E. Psomiadou and M. Tsimidou, *J. Agric. Food Chem.*, 2002, **50**, 716.
2. S. Wang, X. Li, R. Rodrigues, and D. Flynn, *Packaging influences on olive oil quality: A review of the literature*, http://olivecenter.ucdavis.edu/research/files/packaging-influences-on-olive-oil-quality, accessed July 2016.
3. E. Psomiadou and M. Tsimidou, *J. Agric. Food Chem.*, 2002, **50**, 722.
4. T. Mueller, *Extra Virginity: The Sublime and Scandalous World of Olive Oil*, W. W. Norton & Company, 2011.

5. N. Tena, S. C. Wang, R. Aparicio-Ruiz, D. L. García-González and R. Aparicio, *J. Agric. Food Chem.*, 2015, **63**, 4509.

6. R. Mailer, J. Ayton and K. Graham, *J. Am. Oil Chem. Soc.*, 2010, **87**, 877.

7. R. J. Reina, K. D. White and E. G. Jahngen, *J. AOAC Int.*, 1996, **80**, 1272.

8. *Discussion Paper on the Revision for the Limit for Campesterol in the Codex Standard for Olive Oils and Olive Pomace Oils*, ftp://ftp.fao.org/codex/Meetings/ccfo/ccfo24/fo24_13e.pdf, accessed February 2015.

9. P. Fernandes and J. M. Cabral, *Bioresour. Technol.*, 2007, **98**, 2335.

10. M. M. Özcan, *J. Food Process. Technol.*, 2011, **2**, 117.

11. R. Aparicio-Ruiz, M. Roca and B. Gandul-Rojas, *J. Agric. Food Chem.*, 2012, **60**, 7040.

12. R. N. Aparicio-Ruiz, R. Aparicio and D. L. Garci a-Gonza lez, *J. Agric. Food Chem.*, 2014, **62**, 554.

13. R. Mateos, J. L. Espartero, M. Trujillo, J. J. Rios, M. León-Camacho, F. Alcudia and A. Cert, *J. Agric. Food Chem.*, 2001, **49**, 2185.

14. P. Dais, A. Spyros, S. Christophoridou, E. Hatzakis, G. Fragaki, A. Agiomyrgianaki, E. Salivaras, G. Siragakis, D. Daskalaki, M. Tasioula-Margari and M. Brenes, *J. Agric. Food Chem.*, 2007, **55**, 577.

15. E. Karkoula, A. Skantzari, E. Melliou and P. Magiatis, *J. Agric. Food Chem.*, 2012, **60**, 11696.

16. E. Karkoula, A. Skantzari, E. Melliou and P. Magiatis, *J. Agric. Food Chem.*, 2014, **62**, 600.

17. S. Tahmasebi and Z. Rabiei, *The Mediterranean Genetic Code – Grapevine and Olive*, 2013.

18. S. Doveri, D. M. O'Sullivan and D. Lee, *J. Agric. Food Chem.*, 2006, **54**, 9221.

19. F. Cruz, I. Julca, J. Gómez-Garrido, D. Loska, M. Marcet-Houben, E. Cano, B. Galán, L. Frias, P. Ribeca, S. Derdak, M. Gut, M. Sánchez-Fernández, J. L. García, I. G. Gut, P. Vargas, T. S. Alioto and T. Gabaldón, *GigaScience*, 2016, **5**, 29.

20. *Law and Olive Oil: The New Green Gold*, http://www.jurist.org/forum/2012/08/virginia-keyder-olive-oil.php, accessed July 2016.

21. *Standard for Olive Oils and Olive Pomace Oils*, http://www.fao.org/fao-who-codexalimentarius/sh-proxy/en/?lnk=1&url=httts%253A%252F%252Fworkspace.fao.org%252Fsites%252Fcodex%252FStandards%252FCODEX%2BSTAN%2B33-1981%252FCXS_033e_2015.pdf, accessed July 2016.

22. V. Brown Keyder, *Personal Communication*, Aug. 2014.

23. *Trade Standard Applying to Olive Oils and Olive Pomace Oils*, http://www.internationaloliveoil.org/documents/viewfile/9708-norma-english, accessed July 2016.

24. *United States Standards for Grades of Olive Oil and Olive-Pomace Oil*, http://www.ams.usda.gov/AMSv1.0/getfile?dDocName=STELDEV3011889, accessed January 2014.

25. *Evaluation of Olive Oil Sold to Restaurants and Foodservice*, http://olivecenter.ucdavis.edu/research/files/Restaurantsand-Foodservice.pdf.

26. R. Aparicio, L. Roda, M. A. Albi and F. Gutiérrez, *J. Agric. Food Chem.*, 1999, **47**, 4150.

27. A. Carrasco-Pancorbo, L. Cerretani, A. Bendini, A. Segura-Carretero, M. Del Carlo, T. Gallina-Toschi, G. Lercker, D. Compagnone and A. Fernández-Gutiérrez, *J. Agric. Food Chem.*, 2005, **53**, 8918.

28. A. Bendini, L. Cerretani, S. Vecchi, A. Carrasco-Pancorbo and G. Lercker, *J. Agric. Food Chem.*, 2006, **54**, 4880.

29. G. Ciafardini and B. A. Zullo, *Int. J. Food Microbiol.*, 2002, **75**, 111.

30. G. Ciafardini, B. A. Zullo and A. Iride, *Food Microbiol.*, 2006, **23**, 60.

CHAPTER 7

Good Taste is Required

All certified extra virgin olive oils undergo a sensory analysis by
a trained panel of tasters in addition to all of the chemical tests
already outlined. These panels must find the oil to be free of
defects and to possess at least one positive attribute. Who are the
panelists, and how do they evaluate the oils? How can consumers
be sure to select an oil of high quality that they and their families
will enjoy? What does it mean to be a gold or silver award winning oil? In this chapter, we will try to answer these questions,
give you some tools to do your own tastings, and explain how our
sensory apparatus works.

7.1 WHY NOT USE CHEMICAL TESTS FOR TASTE?

The standard chemical tests outlined in Chapter 6 are state of the
art research tests, done by scientists in olive oil chemistry. However,
the tests for free fatty acids, peroxides, or most other quality indices
do not address taste. These are important tests that answer important questions, but they fail to answer the crucial one question: what
does the oil taste like? Despite being an active area of research for
a number of years, chemical analyses for taste, or "electronic noses
[and] tongues"[1] are not yet capable of replacing human tasters. The
problem is well addressed by Aparicio and colleagues:[2]

The Chemical Story of Olive Oil: From Grove to Table
By Richard Blatchly, Zeynep Delen Nircan and Patricia O'Hara
© Richard Blatchly, Zeynep Delen Nircan and Patricia O'Hara, 2017
Published by the Royal Society of Chemistry, www.rsc.org

"... [T]he most updated methodologies of flavor analysis still fail to reproduce the same information provided by sensory assessment. Several reasons explain the gap between instrumental analysis and sensory assessment. The first reason is the high complexity of volatile compounds in virgin olive oil and many other foods, which constitutes a resolution challenge for the current chromatographic techniques. The second reason is the kinetic component of the flavor release occurring during eating and swallowing and influenced by many factors, such as saliva composition and mouth movements. These factors modulate the final sensory perception, and they are not taken into account in most of methodologies for flavor analysis. Finally, little is known about the physiological mechanisms by which flavor compounds result in neural activation and ultimately give rise to sensory perception."

As we describe your sensory apparatus, even in the most simple terms, we hope you begin to realize what a powerful analytical apparatus you are. A good gas chromatograph can separate up to 500 to 1000 different substances in a single run. That sounds quite amazing until you realize that your nose can distinguish at least *10 million and likely over 1 trillion aromas*,[3] without you realizing that you are working at it. While artificial "noses" can detect minute quantities of aroma molecules, extracting the aroma profile means both knowing how the human body responds to each molecule and then how the brain reconstructs the stimulus to produce an aroma. With so many aroma compounds and so many places for them to be detected, we have not yet completely solved that puzzle. Similarly, liquid chromatography can separate a hundred or so compounds, but not tell us the strength of the bitterness response. As it turns out, these are devilish compounds that tend to morph from one form to another, so the usual prospect of testing single compounds one-by-one for bitterness becomes much harder. Let's explore our chemical senses more thoroughly, and as we do, explain the strange sounding (but very logical) actions of the taster.

The first impression one gets of an official tasting room, such as the one shown in Figure 7.1, is a sense of overwhelming blandness. Light colored walls with little ornamentation surround the kinds of study nooks where a student might take an exam or be

Figure 7.1 Tasting room set up for a California sensory panel at an official tasting. From Paul Vossen and A Devarenne, UC Cooperative Extension sensory analysis panel enhances the quality of California olive oil, *California Agriculture* **65**(1), 8–13, Copyright © 2011 The Regents of the University of California. Used by permission.

sent after school if tardy. The only spots of deep color are the cobalt-blue tasting glasses stacked in a cabinet along the wall or arrayed on a heater in readiness for a tasting. The spaces are quiet, clean, and dull. What makes sensory analysis come alive is, of course, the people involved. The leaders of tasting panels we have met are devoted, detail-oriented, and disciplined individuals with extremely discriminating palates. It is the leader's job to train a group of panelists to provide feedback that is statistically consistent and reproducible. The panel must perform well in blind tastings of standards sent out by the certification organization in order to become and maintain active certification. The tasters themselves are passionate members of the community with extensive training who learn to disregard their initial emotional reaction to an oil and dissect their sensory response into dozens of criteria. These individuals must also have a great memory for oils they have tasted in the past and produce similar analyses for the same oil tasted at different times. Perhaps not surprisingly, it can take years to be trained. This is made more amazing by the fact that, if the panel is to be accredited by certain oversight organizations (such as the IOC), the panelists must be volunteers.

One might also ask, "Why a panel?" Sensory analysis depends on both the state of the oil and of the taster. To remove the influence of a single individual who might be having a bad day, panels of 8–12 individuals are convened. Each person analyzes the same group of oils and scores the oils on a number of criteria, generally using a 10 or 15 point scale (you can find some sample tasting score sheets at http://worldolivepress.blogspot.com). The oils are rated on a number of generally positive qualities (such as fruitiness, grassiness, aroma intensity, bitterness or pungency) or negative qualities (such as off-aromas – fusty, muddy sediment, musty, cooked, rancid, and many more). The scores on these qualities must all agree within an appropriate margin of error or the oil is retested.

This may seem like a lot of work, and it is. But here is the essence of extra virgin olive oil: failure to have a single positive quality, or the presence of a single negative quality, demotes the oil from extra virgin to virgin oil or worse. Even if the lab tests tell us that the oil has passed, if it doesn't taste like extra virgin oil to a trained panel, it isn't. This is serious business and fortunes, honors, and reputations can be at stake.

When all is said and done, tasters stand between those who hope to make money from the sale of the product and the consumers. The IOC defines a taster as a "perspicacious, sensitive person who is selected and trained to evaluate the organoleptic attributes of a food with the sense organs." When you, the consumer, have finally made a considered selection of extra virgin oil at a shop, or reach for a bottle of oil in a fine restaurant, you should expect the best. These tasters literally stand between you and bad oil. Every taster, even in international competitions, has had the experience of tasting oils that passed all the chemical tests, and claimed to be extra virgin, but were not. Leaders of tasting panels have told us stories of the delicate phone calls they often must make to inform producers of defects in their oil that will prevent them from gaining the coveted extra virgin label. Trying to be helpful, they often know exactly what caused the defect – and suggest that perhaps the producer should drain the racking tanks *just* a bit more often, or clean the decanter more thoroughly if left overnight, or stop storing diesel fuel in the processing plant. Yes, experts can detect all those defects in the oil, and more.

In fact, a single person can often pick out these defects (or the absence of them) and therefore be an extremely valuable asset to the producer. Some accredited labs offer the option to have oil tested by a single trained person before going to a whole panel. If the oil is to be certified extra virgin, it needs the panel, but to be sold locally, perhaps it just needs a single person to give tasting notes.

7.2   HOW DO THE PROFESSIONALS EVALUATE OLIVE OIL?

First, we must remind you that what follows is not intended as a manual for expert tasting, but rather a generalization of some of the processes experts use.[4]

Cameo 7.1   Margaret Edwards, Matiatia Groves, Waiheke Island, New Zealand.

Margaret Edwards, co-owner of Matiatia Groves in Waiheke Island, New Zealand, was raised in a family in which tasting and appreciating food and discussing flavors were part of each dinner time. She has held numerous positions in her life including faculty member in nutrition science, manager in a food industry, and then in 1997 she and her husband John became producers of olive oil from a small olive grove they planted with their own hands. Margaret's tasting credentials speak for themselves. After being certified by IOOC as a supervisor for tasting panels, she established an accredited tasting panel in New Zealand in 2004. Today she is New Zealand's only international olive oil judge. She lectures all over the world, has written extensively about New Zealand olive oil, and runs tasting workshops. In our trip to their grove, we were impressed by the spotless olive press and comforted by the shared excitement for olive oil quality.

7.2.1 First: Eliminate Distractions

The bland room described above is deliberate. Deep color can be suggestive to other sensations, so is avoided. The room should be quiet, odor free, and unhurried. The senses are especially acute in the morning, so that's a good time to start. The analysts must also prepare themselves. The body, especially the hands, must be free of perfumes (hand creams, body spray, *etc.*). Similarly, the mouth must be free of foreign flavors, as from strong foods, gum, certainly cigarettes.[5] Selin Ertür, producer of Selatin Olive Oils and expert olive oil taster with tasting certification from the Tuscany region of Italy, told us that, in some panels she has worked with, smokers aren't even allowed to be tasters. This makes sense as some smoking can have a negative effect on taste buds.[6]

7.2.2 Prepare the Oils

It's likely that the oils will have been prepared by the panel leader before the other members arrive. It is, of course, vital that no one doing the analysis knows the identity of the oil until the oil has been evaluated and final scores tabulated.

The oils are distributed in dark colored, often cobalt blue glasses, as seen in Figure 7.2.[7] These glasses are placed on heaters designed to gently heat the oil to 28 °C (about 82 °F, and about the maximum temperature recommended for processing). They are covered with glass lids to capture all the developing aromas. The colored glass prevents the color of the oil from distracting the taster and influencing the other sensations of the oil.

7.2.3 Smell the Oil

According to the IOC standards,[4] the first step is to maximize the surface area of the liquid by rolling the glass, keeping it carefully covered. Then, the lid is lifted when the glass is near the nose and a slow, deep inhalation takes in the first contact with the oil. An expert might classify this method of smelling as orthonasal olfaction. This is all about the volatile compounds that we'll discuss below, responsible for most of the smell of the oil. This process should not take longer than half a minute. If needed, the oil can be retested, but only after a brief break.

Figure 7.2   Professional cobalt blue olive oil tasting glass, shown on left, hides the color of the golden olive oil, shown in decanter on right. This is done so that tasters are not biased by the color in their sensory evaluation.

The score sheets suggest that this is a very active 30 seconds! Not only the basic characteristics of fruitiness (green or ripe) should be detected, but also the presence of aromas of tomato, avocado, walnuts, butter, and artichoke (to name just a few). These would not be contaminants, but aromas that naturally occur in the oil itself. Of course, there are also the roughly 12 main defects. Many of these defects and attributes must be scored on intensity scales of 0–10 or 0–15. Experts train by having standards to smell in which a particular positive or negative aroma note is highlighted. Certified tasting labs guard these standards like precious metals, as it is very, very difficult, but not impossible, to re-create them from scratch in the laboratory.

7.2.4   Get the Sensation of the Oil in Your Mouth

After all the aroma notes are made, approximately 3 mL of the oil (about a tenth of an ounce) is brought into the mouth and distributed as completely as possible. The analyst distributes the oil side to side quickly, and more slowly front to back. While doing this, short sharp breaths are taken with the mouth partly open, both to move the oil from front to back and to note any new aromas that occur from the back of the nose (the so-called retronasal aromas). The nose is connected to the throat at the back of the mouth, and aromas perceived there can be different from the front of the nose.

There are two main sensations in the mouth: general mouth feel (thick? thin? astringent or drying?) and bitterness (tasted throughout the mouth, but more intensely in the back). The requirement that the oil be moved to the back of the throat makes swallowing a logical, but not necessary, next step. Some oil tasters, like wine tasters, spit the tasted oil into a waste cup. One of the final sensations is irritating, and can cause a cough or slight burning sensation in the throat. In English, this is called "pungent," which seems a rather generic word. In French, it is called "piquant" (as one would refer to a hot pepper), and in Turkish a similar word is used (yakıcı meaning burning). In truth, there is no perfect word for a nearly unique sensation whose only food source is olive oil. For those who are really curious, the same pungent sensation can be recreated by sampling a suspension of ibuprofen in water in the same manner.

7.2.5 Clear Your Palate and Rest Between Tests

After recording the flavors and intensities for a particular sample of oil, panelists must move on to the second, third, fourth, and so on samples with a fresh mouth. Green apples (such as Granny Smith) will help cleanse the palate, as will rinsing with room temperature water or carbonated water. It is expected that the tasters be given a 15 minute rest between oils. It is critical that tasters follow the exact same protocol for every sample provided to them so as to remove any sampling bias. Sometimes, panel leaders will switch up the order of tasting among the panelists so that one oil is not always sampled first or last.

7.2.6 Collect the Score Sheets and Tally Them

At the end of the session, the panel leader collects the score sheets and looks over the results for any obvious anomalies. If one taster's scores are quite different from the rest, that taster might be invited to review their scores or retest the oils.

Once a generally consistent set is obtained, the scores are tabulated. For positive attributes, all scores are entered. The values must be within 20% of each other by an exact measure specified by the IOC.[4] For negative attributes, a few are always tabulated (such as muddy sediment and rancid) and some are numerically scored if at least 50% of the tasters note that defect. Once scored, the median value is reported (this is the value in the exact middle of a sorted list of scores). If the median value for ANY defect is above zero, the oil cannot be classified as extra virgin.

7.2.7 Judging for Competitions

Only a few of the thousands of trained tasters worldwide will rise to the level of competition judges. These are often the tasting panel leaders and those who have decades of experience with sensory analysis of oils. Competitions occur every year all over the world, and if an oil wins a gold medal or takes best of show or best of class, it can make all the difference to the marketing of that oil. It is critical that a judge not be biased or base a decision on country of origin or cultivar. Blind tasting is the norm so as to minimize these factors, and typically northern hemisphere oils are judged in a separate category than southern hemisphere oils as they are produced six months apart and one will always be six months fresher than the other. Still, it can be subjective which oil wins a gold medal or best of show on a particular day. Sue Langstaff, Owner of Applied Sensory LLC and Olive Oil Sensory Panel Leader, has many years' experience judging extra virgin olive oils. Sue told our UC Davis class on the Sensory Analysis of Olive Oil that these competitions are essentially beauty contests. An oil may win no medal in one competition and take home best of show in another.

Each year, "Flos Olei: guide to the world of extra virgin olive oil" is published by Italian writer and food and wine expert Marco Oreggia, and his colleague Laura Marinelli, journalist and expert olive oil taster.[8] The two coordinate an annual expert tasting of olive oils sent to them from around the world and based on the panel's evaluations, the top 500 oils make it into the guide. The book also provides descriptions and tasting notes for each oil and a score based on a 100 point scale similar to the Robert Parker system for wine.[9] The oils are organized by country, and

Spain and Italy oils are further subdivided by region. The book also provides interesting background information on the culture and history of olive oil in each region. Perhaps it is not surprising that the vast majority of the 20 best extra virgin olive oils listed in "Flos Olei" for 2016 were from Italy and Spain. In *Whose olive oil is really the best in the World?* Aris Kefalogiannis, CEO of Gaea Greek Extra Virgin Olive Oils, famously joked "What we all people in the Mediterranean area believe, whether we are Italians, whether we are Spaniards, or whether we are Greeks, we are absolutely and utterly convinced that the best olive oil in the world is not Greek, Italian or Spanish, it is our village's olive oil!'[10]"

7.3 SOUNDS LIKE A FUN GAME – CAN WE PLAY AT HOME?

Of course! Like many home games meant to represent real life, these results won't be able to withstand a court challenge or grade oil for a competition. However, it's really fun, educational, and allows you to expand your tasting skills and your ability to discuss the finer points of olive oil. "Tasting" an olive oil properly at home will include both smelling (olfactory input) and tasting (gustatory input) and will take about 5–10 minutes per sample, so settle in. Because our noses and taste buds can easily get saturated, only a few oils should be tasted at a time. Like the experts, you can also use a slice of a green apple or water or carbonated water between samples to help cleanse your palate. Combining this with some time with a knowledgeable friend, oil merchant, or a course on sensory analysis will only make it more satisfying.

While choosing the oils you wish to taste will be an obvious prerequisite to a tasting session, as in most activities, it is important to have the right "equipment." For our informal tasting sessions in schools and community venues, it is much less expensive to use 2 oz tasting cups with lids that can be purchased in bulk. These condiment cups can fit nicely several to a standard piece of paper for tracking (we typically use four oils – a sample home tasting sheet is available at http://worldolivepress.blogspot.com). The plastic will detract from the flavor slightly, and the temperature is not optimal, but we have found our students able to taste the differences in oils and appreciate at least some of their unique attributes.

At this point, we will turn you over to Vicki Zancanella from The Olive Press in Sonoma, CA, who has a very relaxed and thoughtful style of tasting oil. Stemming from the rigor of her masters degree in sensory biology, and honed by eight years' experience helping thousands of customers appreciate oils in the tasting room at The Olive Press, her tasting style is very helpful for new tasters.

7.3.1   Vicki's Recommendations for Consumer Tasting of Olive Oil

- Warm the oil: pour approximately one tablespoon of oil into a two-ounce demitasse/condiment cup. Warm it by holding the cup in the palm of your hand while you cover the opening, either with a lid or the palm of your other hand, and gently swirl the oil for a minute.
- Smell the oil: once the oil is warm, take a deep sniff of the oil, exhaling through your mouth. Do this twice. Then suck the aroma in through your mouth, close your mouth, and exhale this time through your nose. Repeat the whole process one more time. This style of smelling the oil integrates sensations from both the front of the nose (orthonasal) and back of the nose (retronasal).

At this point, stop and see if you can describe the smells (*i.e.* green grass, banana, green apple, artichoke, ripe fruits, spices such as cinnamon, cumin *etc.*).

- Taste the oil: next, coat the tip of your tongue and lips by taking a small sip of the warmed oil and hold the oil in the front of your mouth for a few seconds. Note how the oil feels in your mouth (is it thick, thin, astringent?). Don't expect to taste much at this point. Examine that first taste for fruity flavors. Finally, with a larger sip, let the oil coat your entire tongue. You may bring in some air a few seconds later to move the odors up again through your retronasal passage.

What you may notice now is bitter and black pepper flavors on the tongue and a repeat of the aromas in your nose. Most people will swallow the oil and, if pungency is present, you will feel a burn in the back of the throat (you may even cough). While you wait for the pungency to develop (sometimes 30 seconds or more), note how the aroma develops.

This process takes some of the sensations normally observed in parallel by sensory panel tasters and separates them into different steps. It is therefore better suited to people learning to taste, as it allows one to focus on one sensory attribute at a time. Vicki's advice is instrumental in teaching us how to really savor our foods – not just olive oil. While we typically can quickly say that we like or dislike the "taste" of something, our sensory appreciation for a food comes to us by a complex interplay of all of our senses. "Flavor" is a mixture of how the molecules in the food smell – both through the front (orthonasal) and back (retronasal) of our noses, taste through stimulation of the receptors on our tongue, and feel in our mouth (creamy, cold, hot, sticky, flakey, bubbly). Studies show that the sound of the food as we eat it (a crunch of a potato chip) can add to enjoyment or repulsion. Even something as simple as the temperature can affect our appreciation for food. Who wants warm ice cream or cold chicken soup? What parent doesn't know that appearance of food plays a subtle and sometimes misleading role in likes and dislikes? Chefs in high-end restaurants certainly know that when plating food they need to make the food as visually appealing as possible to maximize your enjoyment.

7.4 WHAT ARE WE ALL LOOKING FOR?

In truth, certified tasters, producers, marketers, and consumers are all looking for the same things. We may differ in our ability to discriminate among the many sensations we are bombarded with in a good olive oil, but that is largely a feature of training. Think about the first time you saw a sports event of a sport that is now your favorite (if you can remember). It probably seemed impossibly complex, and you may have thought that you would never know even the most basic rules. Now, you are arguing the fine points of obscure rules and recalling great plays from past games. Tasting olive oil can be just like that.

Ultimately, taste is the most important criterion by which you will decide whether or not you LIKE a particular oil. Having said that, unless you have spent hours doing this as part of your job, or are a certified taster of olive oil, your appreciation of olive oil is probably quite naïve, informed perhaps by your past experiences with olive oil. The Old World consumers, for instance, may find less pungent, sweeter oils more pleasant, associated with memories from the past "good old days." Until recently, most consumers in the USA did not have access to the wonderful array of olive oils prepared all over the world from fresh olives that yielded oil that was properly treated on its way to you. Fortunately, that is changing as we have become aware of the enormous health benefits of certain kinds of olive oils and as retailers have begun to respond to consumer demand for quality and choice. So don't despair, with some very pleasurable homework, your palate can be educated and you will develop your own preferences for oils that will be healthy for you and taste just the way you like.

7.4.1 Aroma Notes

Smell is probably the single most important sense in understanding our appreciation of food in general and olive oil in particular. If you don't believe this, just try to taste anything while holding your nose. What you "taste" in the absence of smell is just a shadow of what is available when your nose gets involved. An olive oil should smell fruity. Exactly which fruits or vegetable smells are there will depend upon the cultivar (type of olive), the terroir (where it was grown), the maturity of the olive when picked, and the age of the oil. A waxy crayon like smell tells you that the oil is rancid. A cucumber flavor suggests the oil was packed too long in a tin container. Write down your impressions right away and don't second guess yourself. First impressions are the best. Keep track of your oil at home and smell it each time before you use it to make sure it is still fresh. Table 7.1 reports the basic positive attributes and negative defects associated with olive oils and shows you some representative types of molecules that can create these aromas.[11-15]

Novices to olive tasting often find it difficult to find the right words to describe the aromas – both good and bad. What we perceive as a "grassy" aroma is likely to be the result of dozens, if not hundreds, of small molecule odorants in different concentrations that together cause our brain to classify a smell as "grassy." Very few people are born with a high level of nasal acuity but it can be trained over time by repeated attention to the fragrances

Table 7.1 Flavor attributes in olive oil and some of the molecules that contribute to these attributes. This is a very short excerpt of a much longer list of contributors to the aromas.

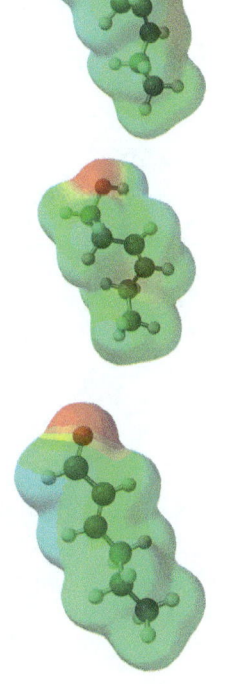

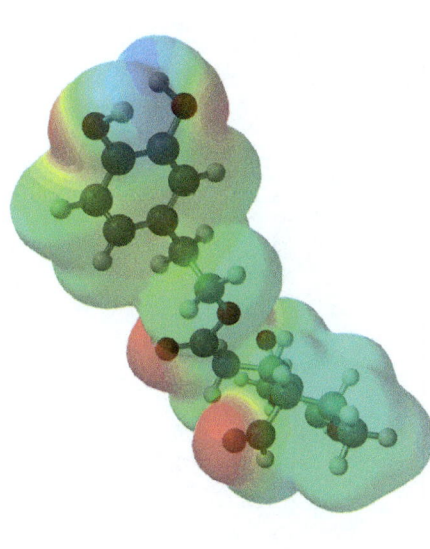

Positive flavor attributes

Fruity   Fruity, as used by tasting panel members, is a familiar smell that comes from both green and black olives if they are fresh, unblemished, and healthy. It includes "green" notes such as grassy or leafy smells (represented here by (*E*) 3-hexenal and (*z*) 3-hexen-1-ol) or "ripe fruit" aromas such as banana, peach, or pineapple (represented here by banana-like *Z*-3-hexenyl acetate).

Bitterness   Bitterness is associated with green olives that have not fully ripened to black. Bitter sensations are felt on the tongue, especially towards the back of the mouth. Antioxidant molecules responsible for many of the health benefits of olive oil are mostly bitter, with varied degrees of bitterness. Hydroxytyrosol, one of the parent compounds, is actually itself not noticeably bitter, but slightly salty. The most bitter compound in olive oil is the compound shown here, DHPEA-EDA.

(continued)

Table 7.1 *(continued)*

Pungency	Pungency is a quality of food that causes a tactile burning at the back of your tongue, just at the entrance of your throat. It will often cause a cough when a taster samples a particularly pungent olive oil. Oleocanthal, the molecule responsible for this burning, and shown here, has a biological similarity to ibuprofen, which also shows pungency. 

Some negative flavor attributes or defects[a]

Fusty	Fusty tastes develop when harvested olives are stored in sacks or containers for long periods of time before pressing. Exposed to heat and light, they will be prone to anaerobic fermentation. Three examples of compounds that give rise to the fusty defect: guaicol, 2-methyl-1-butanol, and butanoic acid, are shown here. 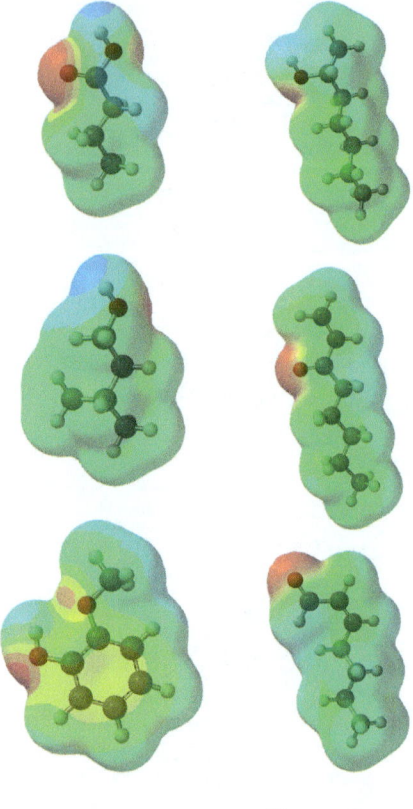
Musty	Musty flavors arise from fruit that has been contaminated with fungus cultivated if the olives are wet or stored for extended periods of time in humid conditions. 1-Octen-3-ol for mustiness–humidity. Three examples of compounds that give rise to the musty defect: (*E*) 2-heptenal, 1-octen-3-one, and heptan-2-ol, are shown here.

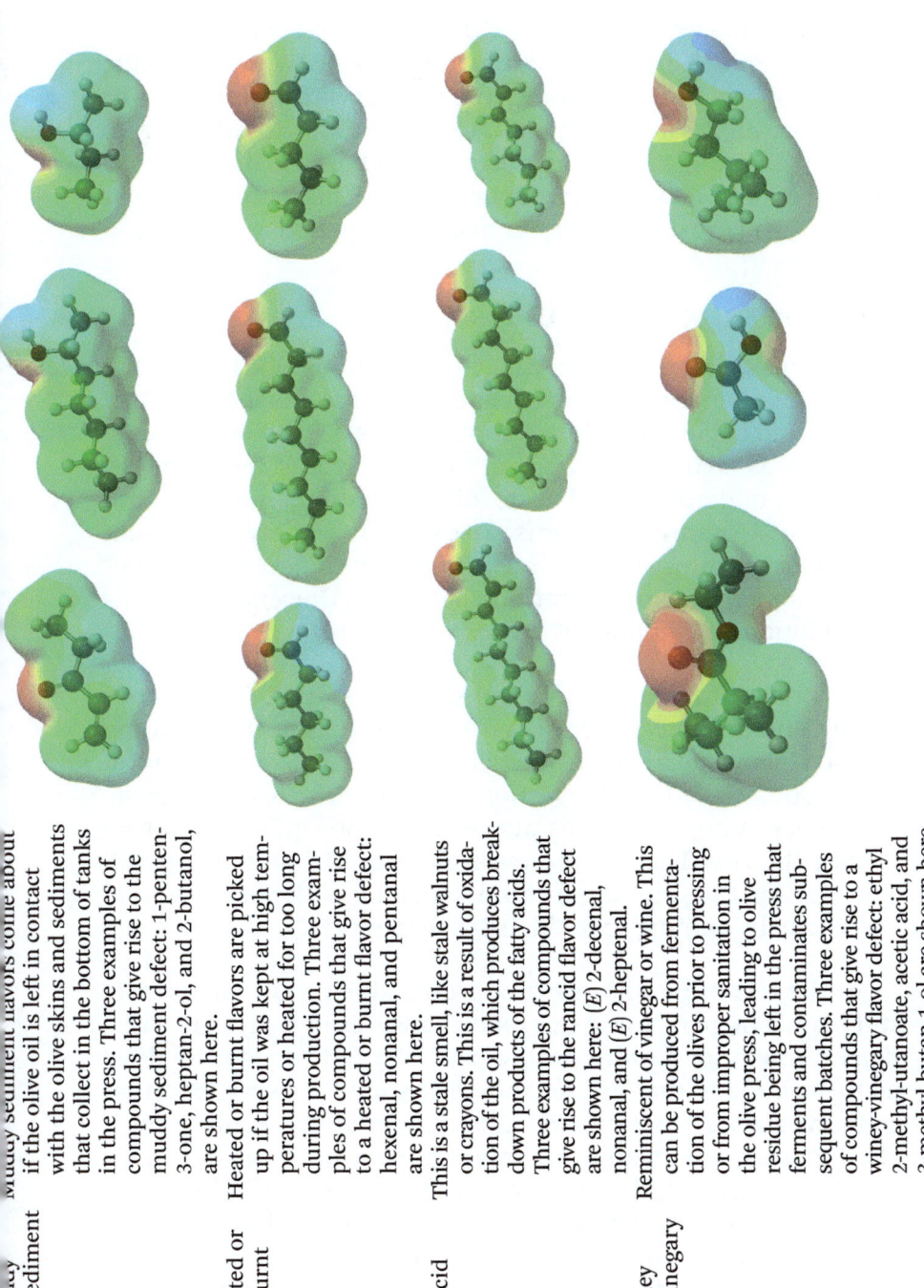

Muddy sediment — Muddy sediment flavors come about if the olive oil is left in contact with the olive skins and sediments that collect in the bottom of tanks in the press. Three examples of compounds that give rise to the muddy sediment defect: 1-penten-3-one, heptan-2-ol, and 2-butanol, are shown here.

Heated or burnt — Heated or burnt flavors are picked up if the oil was kept at high temperatures or heated for too long during production. Three examples of compounds that give rise to a heated or burnt flavor defect: hexenal, nonanal, and pentanal are shown here.

Rancid — This is a stale smell, like stale walnuts or crayons. This is a result of oxidation of the oil, which produces breakdown products of the fatty acids. Three examples of compounds that give rise to the rancid flavor defect are shown here: (*E*) 2-decenal, nonanal, and (*E*) 2-heptenal.

Winey vinegary — Reminiscent of vinegar or wine. This can be produced from fermentation of the olives prior to pressing or from improper sanitation in the olive press, leading to olive residue being left in the press that ferments and contaminates subsequent batches. Three examples of compounds that give rise to a winey-vinegary flavor defect: ethyl 2-methyl-utanoate, acetic acid, and 3-methyl-butan-1-ol are shown here.

[a]These defects, their causes, and aroma characteristics have been nicely summarized in a "Defects Wheel for Olive Oil," © Susan Langstaff (www.defectswheel.com).

and finding the words to share with others a name for what you are smelling. Closing your eyes, taking one or two deep sniffs, writing down your impressions, and talking with your friends or retailer are all great ways to start. Conversation with others gets your language aligned to your perception. Once you develop your vocabulary and can link particular sensory inputs to particular qualities, you are on your way to being a proper olive oil taster.

7.5 EVOLUTION OF HUMAN TASTE

Why is it necessary to have the capacity to taste our foods? Wouldn't it be a lot easier and more efficient to consume a diet focused on calories and balance and forget about the taste? Unlike some species, such as koala bears, who eat only eucalyptus leaves and so have very underdeveloped taste discrimination, humans have evolved to be omnivores; that is, we eat a wide variety of different foods. Our gustatory systems are extremely well developed. Why? Paul Breslin, Professor of Nutrition at Rutgers University, explains the evolutionary component to human tasting.[16]

> "The risks of making poor food selections when foraging not only entail wasted energy and metabolic harm from eating foods of low nutrient and energy content, but also the harmful and potentially lethal ingestion of toxins. The learned consequences of ingested foods may subsequently guide our future food choices. The evolved taste abilities of humans are still useful for the one billion humans living with very low food security by helping them identify nutrients."

For the rest of us with fewer concerns about food safety, taste still plays a huge factor in our food choices even if they are a vestigial remnant of a more hostile past. Not only does taste drive us to seek foods that are nutritionally sound – sweet food that contains sugars, savory foods that contain amino acids – but it also helps us avoid plants that might be toxic or too acidic. Bitter flavors are associated with plant alkaloids that can kill you and sour organic acid flavors often mean rancid or spoiled food. That today we seek to find foods that stimulate these same receptors (coffee, chocolate, and olive oil), and marvel at their health giving properties, is a remarkable little irony of evolution.

Understanding why taste causes us to like certain foods and avoid others is closely linked to understanding our appetite. Our appetite can be triggered even before food enters our mouths through our nose's olfactory system. Smelling food gets us ready to digest foods, by elevating insulin levels in the blood. Enzymes in the saliva or tongue will start the digestion of nutrient carbohydrates (amylases) and proteins (peptidases), and fats (lipases) as soon as the food enters the mouth. This way, the body will be ready to harvest the nutrient content of the food just as it arrives. Our appetites are also closely controlled by taste-like receptors in the upper digestive system that send signals to the brain so that our bodies can adjust our appetites to seek out the foods we need and avoid those we don't need.[17] Bitter taste-like receptors have been identified at the exit of the stomach that stimulate appetite in the presence of bitter compounds in the short term by causing the release of the hunger hormone ghrelin, which will be discussed in more detail in Chapter 8.[18] Perhaps this is why so many human cultures include bitter tasting aperitifs or appetizers in their meal preparations. In the long term, these same signals work to suppress appetite by another mechanism – delaying gastric emptying, thereby prolonging a full sensation. Perhaps this works to prevent continued consumption of the bitter foods that might contain toxins – and can thus be linked to a learned avoidance response. The integration of signals from the eyes, nose, and mouth with those from the downstream digestive regulators works to create a homeostasis in which, when we are in balance, we eat what our bodies need and stop eating when we have had enough.

7.6 QUICK PRIMER ON MAKING MOLECULAR SENSE OF YOUR FIVE SENSES

We experience the world through interactions of our sensory organs – our mouth, our skin, our ears, our nose, and our eyes, as shown in Figure 7.3 – with stimuli from our environment. We taste molecules that enter our mouth, touch objects to understand their hardness or softness, listen to sounds and music through pressure waves that enter our ears, smell aromas that waft through the air to our noses, and detect photons of light that allow our eyes to relay images to our brain. In the following

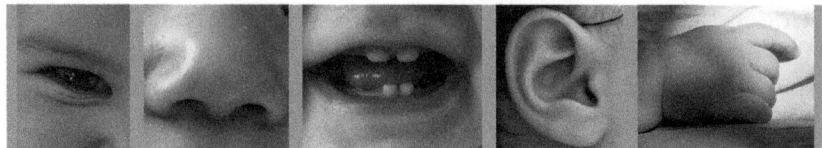

Figure 7.3   Representations of the five senses: taste, touch, hearing, smelling, and vision.

section, we will briefly review how each of the sensory organs
works on a molecular level – with particular attention to the taste
and aromas that let us enjoy good olive oil.

7.6.1   Vision

*"The most pathetic person in the world is someone who has sight
but no vision." — Helen Keller*[19]

A good place to start our discussion of the senses is vision; it
is well-known and makes an excellent model for aspects of the
rest of our senses. Physicists tell us that light travels into your
eye from the outside world through your pupil and is focused
by the lens through the vitreous humor to your retina that lines
the back of your eye. Your retina contains specialized cells
called rods and cones, connected to nerves that feed into the
optic nerve.

The chemical magic begins before the appearance of a photon
– or particle of light. The rod and cone cells in the retina contain
many stacks of receptor proteins called opsin, all of which can
bind to an accessory molecule called retinal (made from vitamin
A). The opsin protein is transparent to visible light, while the retinal is a light yellow. When combined properly, the two create
rhodopsin, a deeply colored protein tuned to act as an antenna
to absorb one of four ranges of visible light. Each cell type has
its own version of the opsin protein and, therefore, responds
to only one of the ranges of light. Rhodopsin can absorb only
one photon of light at a time. When the photon is absorbed,
the bound retinal changes its shape in a staggeringly short time
causing a change in the shape of the much larger protein, which
in turn causes a nerve signal to be sent by the optic nerve to the
visual processing center in the back of your brain. This entire

process takes a tiny fraction of a second. Once used, the retinal – its shape changed – no longer fits snugly in the protein and is removed to be transported to the liver where it is converted back to the active shape. Though slightly different opsin proteins insure perception in each of the four color ranges (blue, green, red, and a general broader absorbance used in night vision), they all work identically. The three different color pigments found in the cones create three different sensors: one for red light, a second for green light, and a third for blue light. The sensory range for the three systems is overlapping, such that a photon of purple light might stimulate either a red receptor or a blue receptor. Our brain will integrate signals from the four visual receptors (one from the rod cells and the three different receptors in the cone cells) to produce a visual image.

If you over-stimulate your receptors, you may experience a (hopefully) temporary blind spot or saturation in which your receptor proteins are bleached and you will literally have a spot on your retina where you will be unable to see. Over time, the retinal migrates back into the eye and rebinds to the opsin and vision is restored. In unusual cases, the retina can be permanently damaged and partial blindness will become permanent. As we will see, a similar thing can happen with your taste receptors.

All senses work in a similar fashion: some input from the environment is detected by a sensory apparatus and communicated to the brain. In vision the input is a photon but in taste and smell, it is a small molecule that might change the shape of a specific protein receptor or disrupt the electrical balance across a cell membrane and trigger a nerve signal to be sent to the brain.

With olive oil, the first thing many of us will notice is the color, which can range from an emerald green to a golden orange–yellow. Many people associate a green oil with a fresh bitter oil produced from olives that are very young. Yellow oils are associated with sweeter oils produced from more mature fruit. While this can be true, the color can also be determined by a host of factors including the cultivar used and the country in which it is grown, the particular processing method, and the post-processing filtering or storage conditions. Because of this, the tasters use the dark colored glass so their tasting of the oil will not be influenced by the color.

7.6.2   Smell

"I wish we could see perfumes as well as smell them. I'm sure they would be very beautiful." – L. M. Montgomery

Smell might be the most underappreciated of the five senses. In all cases, the signal that is perceived in an olfactory response is a small molecule from the environment that is volatile; that is, it must be able to vaporize so that it can diffuse through the air to your nose. Once the small molecule enters the nose, it will travel over the mucous membranes until it reaches an olfactory receptor protein that has a binding site complementary to the small molecule. Once a molecule does bind to a receptor, a signal begins to be composed that will be sent to the olfactory bulb at the base of the brain and then to the brain itself. This small molecule becomes an odorant.

Not all volatile compounds will bind to a receptor, however. Some, like nitrogen, water, or carbon monoxide, do not stimulate any receptor and are odorless, which explains why many homes use carbon monoxide detectors to protect us from deadly levels of this odorless, colorless gas.

Smell is a primal sense. As shown in Figure 7.4, the olfactory neural network connects deep in the brain at such a primitive part that we are often unconscious of its workings.[20] For the most part, we have strong associations of emotions and particular places that are linked with particular smells. No doubt you have had the experience of evoking a memory of a childhood place or person from experiencing a particular smell. As the sense of smell is so primal, it is one of the last senses to go when a person's mind deteriorates. Similarly, individuals with illnesses or disorders who become insensitive to smells can be in real danger in a modern world if they cannot smell a fire or consume a food that is rotten or spoiled.[21]

Unlike the six basic taste receptors in the mouth, genetic analysis has revealed 350 different types of functional odor receptors in the human nose that arise from different olfactory genes scattered throughout our chromosomes. These 350 receptors allow us to recognize millions of aromas.[3] Some animals, such as dogs or mice, have an even more finely developed sense of smell, in part because they have a larger array of receptor proteins.

The sensitivity of our nose – or the lowest concentration of odorants that we can detect – varies with the odorant. If each

Odorant Receptors and the Organization of the Olfactory System

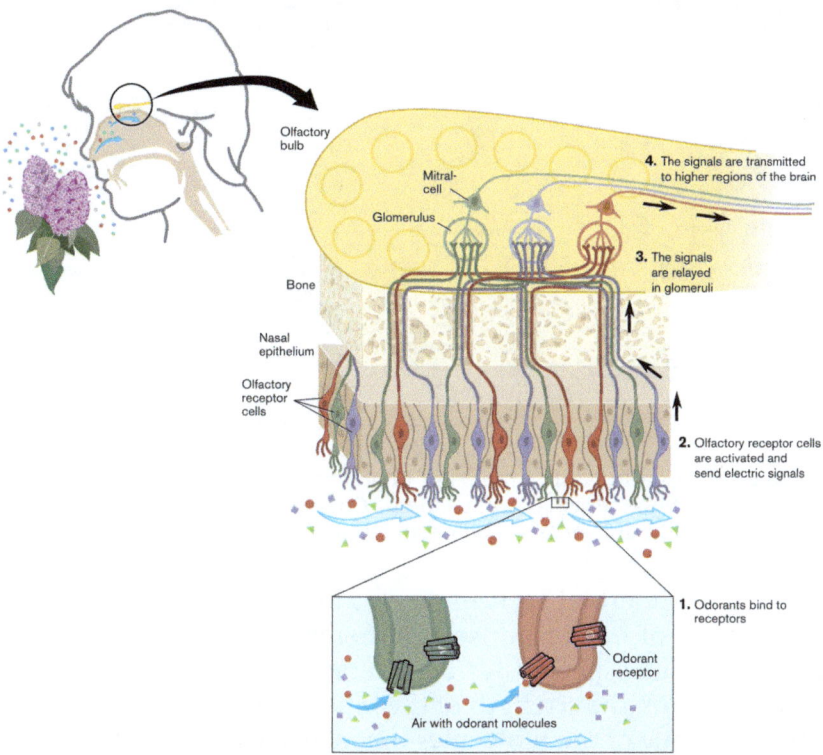

Figure 7.4 Odorant receptors and organization of the olfactory system, from the press release for the 2004 Nobel Prize in Physiology or Medicine to Richard Axel and Linda Buck. Copyright: © Nobel Committee for Physiology or Medicine. Illustrator: Annika Röhl.

odorant binds specifically and selectively to one receptor, how can we distinguish so many molecules when we only have 350 different receptors? In her 2004 Nobel Lecture, Linda Buck explained that the answer is that each odorant might bind to several different types of receptors, and the resultant signal sent to our brains *via* the olfactory bulb is not linear (one odorant–one receptor) but combinatorial, that is, derived from a combination of inputs.[22] As shown in Figure 7.5, a small molecule odorant A binds to receptor type 5 while odorant C, a similar molecule with just one more carbon, binds to a suite of receptors. The signals transmitted to the olfactory bulb for the two odorants, one from receptor 1 and the other from receptors 1, 4, 5, 7, 10, and 11, will also be interpreted as different odors.[23]

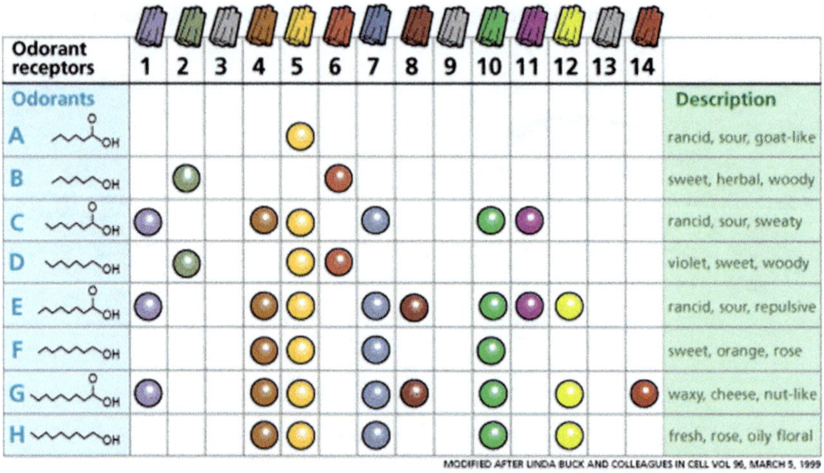

Odorant receptors	1	2	3	4	5	6	7	8	9	10	11	12	13	14	Description
Odorants															Description
A					●										rancid, sour, goat-like
B		●				●									sweet, herbal, woody
C	●				●	●		●			●	●			rancid, sour, sweaty
D		●			●	●									violet, sweet, woody
E	●				●	●		●	●		●	●	●		rancid, sour, repulsive
F					●	●		●			●				sweet, orange, rose
G	●				●	●		●	●		●		●	●	waxy, cheese, nut-like
H					●	●		●			●		●		fresh, rose, oily floral

MODIFIED AFTER LINDA BUCK AND COLLEAGUES IN CELL VOL 96, MARCH 5, 1999

Figure 7.5 Combinatorial model of odorant response, taken from the 2004 Nobel Prize announcement. Copyright: © Nobel Committee for Physiology or Medicine. Illustrator: Annika Röhl.

The combinatorial logic of smell can also explain two important aspects of our response to different amounts of odorants. First of all, as expected for an interaction between a molecule and a receptor, the strength of the binding depends on the structures, so some of the molecules that activate a receptor will bind strongly and some weakly. Molecules that bind very strongly will activate the receptor at lower concentrations, and we would say the odor threshold concentration is lower (that is, we can smell smaller amounts of these). More weakly binding molecules require higher concentrations to be perceived, so we will measure a higher odor threshold concentration. This also helps to explain one important quirk of our perception of odors: the tendency of some compounds to give a different perception at different concentrations. With a small amount of compound B in the air, perhaps it binds only to receptors 2 and 6. At higher concentrations, it may begin to activate a third or fourth receptor, changing our perception.

For olive oils, the positive attributes are composed of volatile chemicals, shown in Table 7.1, many of which are produced during the processing of the oil. The fruity aromas come from the natural enzymatic breakdown of the TAGs by enzymes in the fruit. In nature, these aromas would have been made naturally as the fruit ripened so as to attract birds to eat the fruit and disperse the seeds. A master miller will know exactly when the best "bouquet" from the oil has been produced in the malaxer. Some odors rise quickly with time in the malaxer and then fall, others take more time to develop, and still others just degrade with time in the malaxer. All of them are sensitive to temperature and time. We have seen many mill managers standing over the olive mash and taking a good whiff to decide if the mash should be moved along to the next phase, separation. Defects in quality of the olives prior to processing can also be picked up here (rancidity, fusty, musty, winey vinegary).

Johan Lundström, a cognitive neuroscientist at Monel Institute in Philadelphia, has shown that humans are capable of distinguishing the fat content of different milks based purely on the smell, a surprising result given the low vapor pressure of the fats.[24] Still, knowing the acute sensitivity of the nose and the enormous evolutionary pressure to choose food with a high caloric value, we should not be surprised. Whether this is due to the fatty acids themselves, or related molecules associated with fat in the milk, remains to be seen. Still, the molecular story is not a complete one. Sharing the 2004 Nobel Prize for olfaction, Richard Axel sums up *"Our genes create only a substrate upon which experience can shape how we perceive the external world."*[25]

7.6.3 Taste

"I have the simplest tastes. I am always satisfied with the best."
– Oscar Wilde

Taste is mediated through three types of taste buds (fungiform, foliate, and circumvallate) that appear in three different areas on your tongue, as shown in Figure 7.6(a). Older text books presented a layout or map of the tongue that illustrated where each taste was located: sweet on the tip, salty on the front sides, sour on the back sides, and bitter in the back. Newer research has shown that this map is not accurate, but that instead, taste receptors of each type are more or less randomly distributed on the tongue with a higher concentration towards the perimeter of the tongue and fewer in the center, as shown in Figure 7.6(b). An exception to this is the bitter receptors, which are in

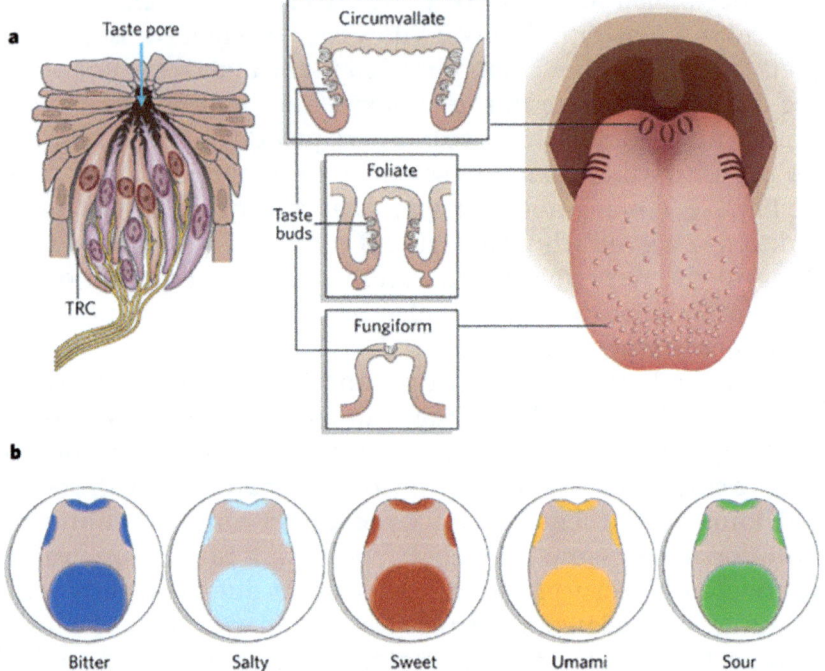

Figure 7.6   (a) Taste is perceived through three types of taste buds on the tongue. Cells within each taste bud responsive to bitter (dark blue), salty (light blue), sweet (red), umami (yellow), sour (green) and fat (not shown) are clustered within each taste bud. (b) In contrast to old taste maps of the tongue, this new map shows that all tastes are perceived by all areas of the tongue. Reprinted by permission from Macmillan Publishers Ltd: *Nature*, **444**, Chandrashekar, J. *et al.*, © 2006.

fact, clustered towards the back of the tongue at the entrance to the throat. It is likely that this was evolution's attempt to prevent ingestion of toxic bitter plants by causing our foraging ancestors to gag and spit out that bitter leaf before being poisoned.

Figure 7.7 outlines the most likely suspects that reside in our taste buds and are responsible for us actually tasting our foods. These are the proteins that sense the flavor molecules, known to scientists as the receptors (meet T1R1, T1R2, T1R3, and T2R), the pores (responsible for intracellular acidification), and the channels (meet ENaC and CD36). Let's take a minute to get to know them better.

Sweet taste receptors bind to many sugars (such as glucose, fructose, lactose, *etc.*, which have superficially similar molecular

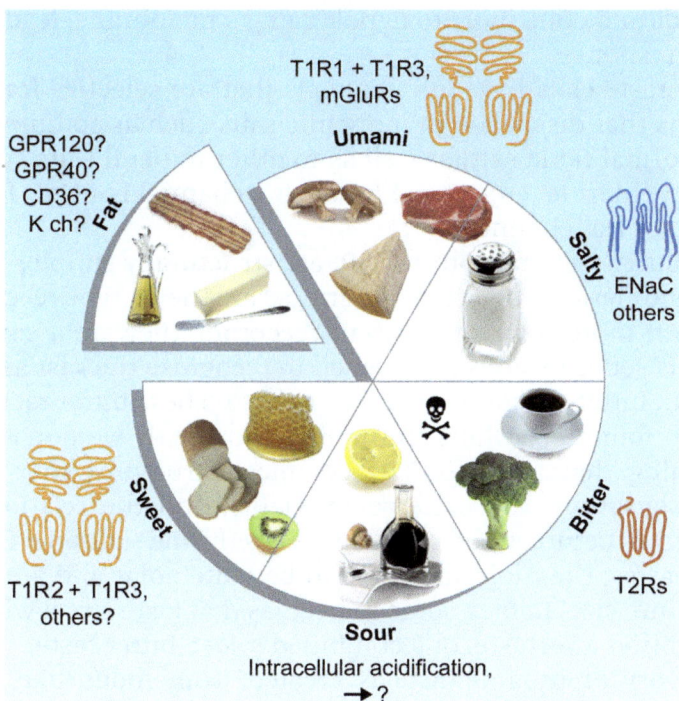

Figure 7.7   Six fundamental taste responses are stimulated by foods containing certain small molecules. Foods containing these tastes are shown towards the center of the pie and the biological molecules that respond are shown towards the outside of the pie. © Chaudhari and Rober, 2010. Originally published in *Journal of Cell Biology*. DOI: 10.1083/jcb.20100314420100927c Rockefeller Press.

structures) as well as some molecules with quite different structures such as chloroform, the amino acid alanine and, of course, a range of artificial sweeteners. The receptors are classified as T1R2 and T1R3 proteins and fall into a larger class of transmembrane signaling proteins called G-Protein Coupled Receptors, or GPCR. Binding to the sugar on the outside of the cell causes a change in the part of the receptor inside the cell, dissociation of a subunit, and a signal is sent. (Sound familiar? – remember vision!) Molecules that give rise to a sweet taste are generally polar and have at least a section of the molecule which has these polar portions arranged in a specific three-dimensional arrangement.

Sour taste arises from the passive diffusion of acids such as hydrochloric, acetic, or citric acid through membranes lining the taste buds. All acids have protons (H^+) that can dissociate from the acids and contribute to depolarizing a membrane, leading to a taste response.

Salty taste cells have ion channels that are selective for positive ions that dissociate in inorganic salts such as sodium chloride (normal table salt) as well as to other mineral salts such as calcium chloride. One name for such a channel is ENaC for the sodium or Na^+ channel.

The bitter taste receptors, T2R, are structurally simpler – they are monomeric rather than dimeric, as are the T1 type receptors. But, with more than 35 different receptors, they are a complex set. This set of receptors is needed to recognize the vast array of different bitter compounds we respond to. These bitter receptors are also found elsewhere in the body, so clearly we don't know everything that they do. The chemical structures that cause bitter sensations are extremely variable, and the relationship between structure and taste is not perfectly understood. Like the sweet tastes, the molecules tend to be quite polar and are often of medium size. In fact, some molecules that taste sweet will also have a bitter aftertaste, or a combined sweet–bitter taste.

A savory or umami taste is created from foods like meat, cheese, or certain vegetables like ripe tomatoes through T1R1 or T1R3 receptors. These receptors, again dimeric transmembrane GPCR proteins, bind to molecules containing salts of glutamic or aspartic acid and mono nucleic acids such as guanyl, inosityl, or adenosyl groups on the outside of the cell. Binding causes dissociation of an internal subunit and a signal is sent that is interpreted by the brain. Some foods – cheese, mushrooms, steaks – contain these two ingredients naturally. However, chefs have learned how to add this savory flavor to any foods by creating a condiment which combines the two ingredients – an amino acid and a nucleic acid – from foods rich in one with a source rich in the other. The resultant mixture creates a savory condiment that enhances the flavor of other dishes so that less salt or fat need be added. This gastronomic trick was used by the Romans in their use of the condiment garum, made from boiling anchovies rich in both nucleotides and glutamate. The Japanese today enhance flavors of many of their dishes by adding "dashi" – made from the seaweed kelp, rich in glutamate, with dried bonita flakes, rich in the mononucleotide inosinate.[26]

Humans have a total of about 2000–4000 taste buds, each of which remains in service for about a week before being replaced. Each taste bud has between 10–50 sensory cells that respond individually to the whole range of sweet, salty, sour, savory, bitter, or fat molecules in our foods. Each sensory cell contains many copies of its particular taste receptors or ion channels. About half of the sensory cells have mixtures of receptors (two or more of the bitter receptors, for example), which allows the system to vary the intensity of the response to the six different tastes and combinations of flavors.

We should note that chemically similar molecules that bind to the same receptor can cause enormously varying responses, as shown in Table 7.2. For example, aspartame and sucrose both bind to the sugar receptors T1R2 and T1R3, but aspartame is perceived as 150 times more sweet than sucrose. Saturation of your taste buds is also an issue. Like your eye, your taste buds can get overwhelmed and food that tasted really sweet at the beginning of a meal may lose its flavor.[27]

Can one taste change another? Certainly! Adding salt to reduce the bitterness of foods is something cooks have done for centuries. A simple experiment can help you to prove this to yourself. First, sample some tonic water, which contains the very bitter compound quinine.[28] Despite the fact that tonic already contains a sweetener, it is still bitter. Then, add some salt and taste

Table 7.2 Relative potency of different substances. Compounds representing various chemical classes can stimulate taste. Sensitivity can vary as much as several thousand fold from one compound to another. Responses are related to the first item in the list and are a measure of the sensitivity relative to that substance.

Sour	Index	Bitter	Index	Sweet	Index	Salty	Index
Hydrochloric acid	1	Quinine	1	Sucrose	1	Sodium chloride	1
Acetic acid	0.55	Caffeine	0.4	Aspartame	150	Sodium fluoride	2
Citric acid	0.46	Nicotine	1.3	Glucose	0.8	Lithium chloride	0.4
Lactic acid	0.85	Denatonium	1000	Alanine	1.3	Ammonium chloride	2.5

it again. You will be amazed at how even a little salt will reduce the bitterness. The molecules that originally tasted bitter are still there and must still interact with your bitter receptors but their response is suppressed by the salt.

Exactly how the basic "tastes" interact with one another is complex at a molecular level, but knowing that a single taste bud can contain receptors for multiple tastes makes this perhaps a little less mysterious. The taste buds are set up for cross-talk between and among the cells, providing an almost limitless palette that expands our tasting repertoire beyond six basic flavor notes.

It is probably not surprising to you that genetic variation exists with respect to our capacity to actually taste different foods. For example, in one group of men undergoing colonoscopy screening, those with hypersensitivity to bitter tastes ate fewer bitter greens and vegetables and had a greater number of polyps – which suggests a higher risk for colon cancer.[29] As mentioned, the bitter taste receptors themselves likely evolved to protect humans from toxic plants in the environment. Yet today, bitter compounds in cruciferous vegetables like broccoli and Brussels sprouts are known cancer preventative compounds. The genetic variation in the human population with regard to the capacity to taste bitter compounds, such as those from these vegetables, makes some humans supertasters. These individuals are extremely sensitive to these flavors and are often repelled by the flavor and reject these healthy foods and thus can be at a nutritional risk.[30] Thus, while these supertasters were better able to avoid toxic foods in our uncertain past, over a lifetime in a modern world, they are at higher risk for colon cancer and will have different health outcomes.

The most prominent perceptions we will get from putting olive oil in our mouth will be bitterness on our tongues, pungency in the throat, and mouth feel associated with the coating and astringency of the oil. Rancid oils can leave a waxy thick sensation on our tongues. Winey-vinegary flavors will taste sour. Retrograde aromas will also stimulate fruity receptors in the back of the nose. These flavors can take time to develop and so one should take time to develop the full taste and aroma profile.

New research shows that our taste buds are also sensitive to fat, and the oleogustus cells that carry that sensitivity have special proteins that have just begun to be identified.[31] While the triglycerides that make up the bulk of any fat or oil are themselves tasteless, free fatty acids can be made from the triglycerides through the action of lingual lipase. Regions on the tongue discriminate the length and saturation level of these free fatty acids.[32] While the oleogustin cells are still not fully understood, a pore protein, CD6, seems to pass free fatty acids from the mouth into the cell to create a cell signal. Human tasters show a preference for foods containing the essential fatty acids linoleic and linolenic over oleic acid.[33] Rick Mattes, Professor of Nutrition at Purdue University, has summarized his evidence for the existence of this newly found "taste" receptor, which is not to be confused with the tactile differences we can detect based on the smoothness or oiliness of our foods.[34] It is also becoming increasingly clear that taste receptors for fat on our tongue can play an important role in regulating appetite.[35]

7.6.4 Touch

"Touch has a memory." — John Keats

The perception of texture in the mouth, or "mouth-feel," is another aspect of food appreciation. Receptors on your tongue are sensitive to the texture of the food you eat. Sticky, smooth, creamy, oily – all these can cause us to appreciate a food or be repelled by it. This will be very important in the appreciation of olive oils. A watery olive oil? A viscous, sticky olive oil that makes your tongue seem to stick to the roof of your mouth? You know that both of these will provide a negative "tasting" attribute to any oil you will evaluate. But a smooth oil that coats the tongue evenly and persists for a few seconds is just right.

Astringency, or a drying of the tongue, is also a perception associated with a tactile or touch sensation and not a taste. Many wines and olive oils will dry out the tongue, which can be a positive (crisp) or negative (drying) sensation depending on the intensity.

7.6.5 Sound

"One person's data is another person's noise." — K. C. Cole

While it is hard to imagine how this sense is relevant to olive oil, we include it to complete our sensory panel of tools that we have to experience the outside world. Sound is also quite a bit different from other senses in that there is no molecule or photon that triggers a signal that is interpreted by the brain. Rather, sound waves carried through the air cause small hairs to vibrate in our inner ear. The vibration itself is resonant with the sound wave – and we hear sounds of different frequencies and loudness.

7.7 WHAT HAPPENS DURING TASTING

Now that you know how the senses work, you can appreciate all the hard work your sensory organs are doing to explore the chemical complexity of olive oil. When you sniff the oil, you are bringing in the smaller, less polar molecules found in the oil (remember how most of the aroma compounds were fairly small). Warming the oil increases the vapor pressure of these molecules, and makes them more concentrated in the air above the oil, and therefore easier to detect. That sniff brings the molecules in contact with the receptors for sensation. Careful not to sniff too often, though, as the receptors can become accommodated, and the sensitivity is lost.

As you taste the oil, you are touching it, and getting a feeling for the thickness and astringency. After coating your tongue with the oil, however, the bitter compounds, such as DHPEA-EDA, are extracted out of the oil into the watery fluid around the taste buds, and we get the sensation of bitterness. This sensation can last quite a while, due perhaps to the binding strength of the receptor or the lower solubility of the bitter compounds in water.

Of course, our brain is the ultimate arbiter of our perception. The sensory signals project onto the various areas of the brain by means of our sensory nervous system. The array of signals is mixed, memories are tapped into, and for odor, our first response is often emotional. YUM! or YUK! The language center in our brain is located toward the back of our brain, near where vision is processed, and is actually quite a distance from where our brain processes smells. Perhaps this is why it is so hard to use words to describe a taste. Still, practice makes perfect. Positive olive oil descriptors used by professional tasters include grassy, herbal, artichoke, green banana, green apple, green tea, tomato leaf, cinnamon, malty, citrus, floral, perfumed, nutty, almond, buttery, and perhaps most simply, olive. Negative descriptors include woody, moldy, dank, brined olive, vinegary, glue, salami, burnt, metallic, earthy, vanilla, old ice cream, vomit, and old baby diapers. We learned that "cat pee" is actually a positive attribute according to the professionals, but we will leave it to you to seek out that particular varietal. The better we can get about using words to describe what we like and don't like in an oil, the better we can be guided to an oil we will enjoy. This takes repetition and it takes time. It is not an impossible task that only a few people can master. Everyone who tries will be rewarded with an improved appreciation for the oil and each time you will get better.

If you are not convinced that your brain knows the difference between a rancid oil, even if you don't have the words for it (yet), Figure 7.8 might convince you. This image is a functional brain

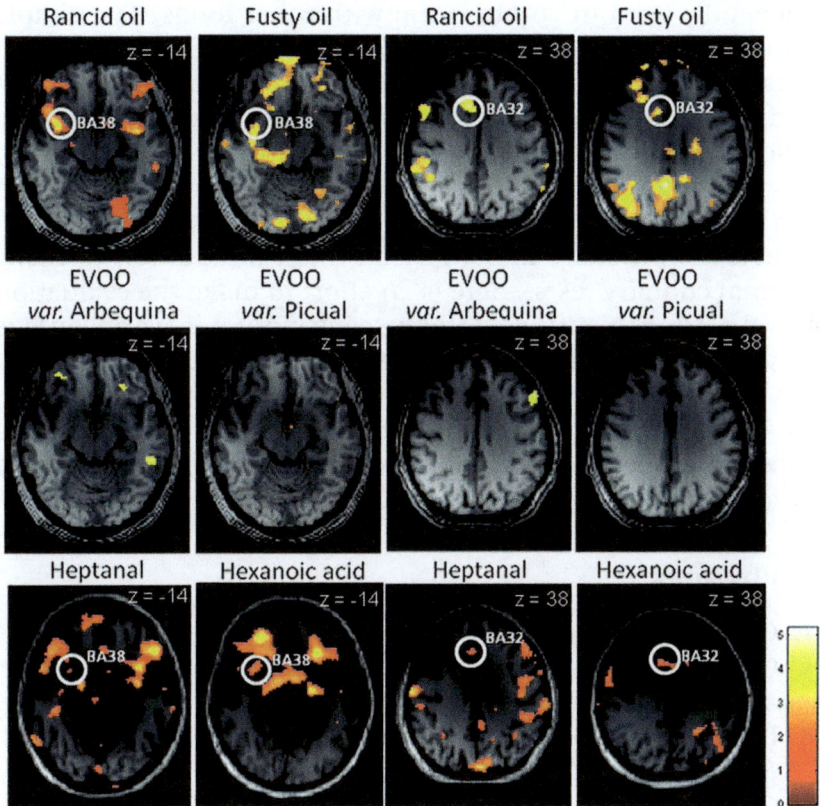

Figure 7.8 Activation responses to the aromas of extra virgin olive oil (EVOO) and virgin olive oils with sensory defects. Reprinted with permission from D. Garcia-Gonzalez, J. Vivancos, R. Aparicio, Mapping brain activity induced by olfaction of virgin olive oil aroma. *J. Agric. Food Chem.*, **59**(18), 10200–10210. Copyright © 2011 American Chemical Society.

MRI from subjects who were exposed to aromas of rancid olive oils and three different extra virgin olive oils. A few things to note are first that most of the brain activity in response to an aroma is NOT in the back of the brain where the language centers are and second, that the rancid oils, and the odorants responsible for the defects, elicited brain activity in a large number of sites as compared to the extra virgin olive oils.

7.8 OLIVE OIL IS THE PERFECT MIXER

Olive oil is like the perfect dinner guest. It doesn't matter where you put it, it will make everything around it taste better. Olive oil is generally used in combination with other foods. The simple mixture of olive oil on bread with a sprinkling of sea salt is delicious, and not quite so bitter due to the salt. Similarly, mixing a good, grassy olive oil with lemon juice makes a delicious dressing for greens, in which the flavors harmonize with the aromas of the greens. Olive oil will provide appetizing fruity aromas when drizzled on warm vegetables, and pair well with chocolates in a dessert. It's probably not surprising that, in addition to many informal culinary tests, there is an effort to make the evaluation of food pairing scientific.[36] Strictly scientific or not, we encourage these experiments and discuss these further in Chapter 9, 1001 Uses for Olive Oil. In addition, olive oil will leave you healthier, by its own nature but also because of the healthy foods it makes more palatable. The next chapter will outline some of the details of exactly how these health benefits work.

REFERENCES

1. C. Apetrei, I. M. Apetrei, S. Villanueva, J. A. De Saja, F. Gutiérrez-Rosales and M. L. Rodriguez-Mendez, *Anal. Chim. Acta*, 2010, **663**, 91.
2. D. L. García-González, J. Vivancos and R. Aparicio, *J. Agric. Food Chem.*, 2011, **59**, 10200.
3. C. Bushdid, M. O. Magnasco, L. B. Vosshall and A. Keller, *Science*, 2014, **343**, 1370.
4. *Sensory Analysis of Olive Oil: Method for the Organoleptic Assessment of Virgin Olive Oil*, http://www.internationaloliveoil.org/documents/viewfile/3685-orga6, http://www.internationaloliveoil.org/estaticos/view/224-testing-methods, accessed March 2016.

5. P. Vossen, 2.1 OLIVE OIL CATEGORY, *International Olive Council (IOC) and California Trade Standards for Olive Oil*, 2007, http://cesonoma.ucanr.edu/files/27262.pdf, accessed December 2016.

6. P. Pavlos, N. Vasilios, A. Antonia, K. Dimitrios, K. Georgios and A. Georgios, *BMC Ear, Nose Throat Disord.*, 2009, **9**, 9.

7. *Sensory Analysis of Olive Oil Standard Glass for Oil Tasting*, www.internationaloliveoil.org/documents/view-file/3669-orga2, accessed July 2016.

8. M. Oreggia and M. Laura, *Flos Olei 2016: A Guide to the World of Extra Virgin Olive Oil*, Marco Oreggia, Roma (IT), 2016.

9. *Wine Tasting Scoring*, https://www.erobertparker.com/info/legend.asp.

10. *Olive Oil And Hot Potato Test*, https://www.youtube.com/watch?v=VTXgLyszsc8, accessed July 2016.

11. J. Reiners and W. Grosch, *J. Agric. Food Chem.*, 1998, **46**, 2754.

12. R. Aparicio and G. Luna, *Eur. J. Lipid Sci. Technol.*, 2002, **104**, 614.

13. C. M. M. Kalua, M. S. S. Allen, D. R. R. Bedgood, A. G. G. Bishop, P. D. D. Prenzler and K. Robards, *Food Chem.*, 2007, **100**, 273.

14. L. Cerretani, M. D. Salvador, A. Bendini and G. Fregapane, *Chemosens. Percept.*, 2008, **1**, 258.

15. G. Dierkes, A. Bongartz, H. Guth and H. Hayen, *J. Agric. Food Chem.*, 2011, **60**, 394.

16. P. A. Breslin, *Curr. Biol.*, 2013, **23**, R409.

17. S. Janssen and I. Depoortere, *Trends Endocrinol. Metab.*, 2013, **24**, 92.

18. S. Janssen, J. Laermans, P. J. Verhulst, T. Thijs, J. Tack and I. Depoortere, *Proc. Natl. Acad. Sci. U. S. A.*, 2011, **108**, 2094.

19. Retrieved from http://www.goodreads.com/quotes/12998-the-most-pathetic-person-in-the-world-is-some-one, accessed July 2016.

20. L. Buck and R. Axel, *Cell*, 1991, **65**, 175.

21. C. P. Kimmelman, *Disorders of Taste and Smell*, American Academy of Otolaryngology–Head and Neck Surgery Foundation, Alexandria, VA, 1996.

22. L. B. Buck, *Angew. Chem., Int. Ed.*, 2005, **44**, 6128.

23. B. Malnic, J. Hirono, T. Sato and L. B. Buck, *Cell*, 1999, **96**, 713.

24. S. Boesveldt and J. N. Lundström, *PLoS One*, 2014, **9**, e85977.

25. R. Axel, *Angew. Chem., Int. Ed. Engl.*, 2005, **44**, 6110.

26. *Umami – The Delicious 5th Taste You Need to Master*, http://www.molecularrecipes.com/molecular-gastronomy/umami/.
27. G. E. DuBois, D. E. Walter, S. S. Schiffman, Z. S. Warwick, B. J. Booth, S. D. Pecore, K. Gibes, B. T. Carr and L. M. Brands, *ACS Symp. Ser. Am. Chem. Soc.*, 1991, 261.
28. K. Chang, *Salt Trumps Bitter*, 2009.
29. B. Garcia-Bailo, C. Toguri, K. M. Eny and A. El-Sohemy, *OMICS: J. Integr. Biol.*, 2009, **13**, 69.
30. S. Reynolds, C. M. Kreider, L. E. Meeley and R. M. Bendixen, *J. Rare Disord.*, 2015, **3**, 1.
31. N. Chaudhari and S. D. Roper, *J. Cell Biol.*, 2010, **190**, 285.
32. R. D. Mattes, *Annu. Rev. Nutr.*, 2009, **29**, 305.
33. C. A. Running and R. D. Mattes, *Am. J. Physiol.: Gastrointest. Liver Physiol.*, 2015, **308**, G442.
34. C. A. Running, B. A. Craig and R. D. Mattes, *Chem. Senses*, 2015, **40**, 507–516.
35. D. Liu, N. Archer, K. Duesing, G. Hannan and R. Keast, *Prog. Lipid Res.*, 2016, **63**, 41.
36. L. Cerretani, G. Biasini, M. Bonoli-carbognin and A. Bendini, *J. Sens. Stud.*, 2007, **22**, 403.

Health Effects: But is Olive Oil *Good* for You?

In August of 1997, Jeanne Louise Calment of Arles, France, died at the ripe old age of 122, as fully documented and recorded in the *Guinness Book of World Records*. What was the secret to her amazing longevity?

"...olive oil, port, and chocolate!"[1]

Mrs Calment may not have known that at least part of her secret formula, olive oil, has been considered vital for a healthy life for at least 2000 years.

Dioscorides, great physician, pharmacologist, and botanist of antiquity advised Romans in the 1st Century how to balance the four temperaments that governed their health (hot, cold, wet, and dry) by using oils produced from olives harvested at different degrees of maturity. Whether to ease constipation, soften skin, or even prevent hair from turning grey, there was an oil to match.[2]

"The characteristic of olive oil corresponds to the type of olives from which it comes. That which is pressed from ripe olives is the

The Chemical Story of Olive Oil: From Grove to Table
By Richard Blatchly, Zeynep Delen Nircan and Patricia O'Hara
© Richard Blatchly, Zeynep Delen Nircan and Patricia O'Hara, 2017
Published by the Royal Society of Chemistry, www.rsc.org

most moderate and the best; oil taken from unripe olives contains cold and dryness. That pressed from red olives is midway between the two other oils. Oil pressed from black olives will warm and moisten moderately, and it is beneficial for poisons, loosens the belly, and expels worms. Oil from old olives has a greater strength to warm and dissolve. That which is pressed out with water is less hot and gentler, more effective and beneficial. All of its types soften the skin and delay whitening of the hair. Liquid of the salty olive prevents blistering of burns and strengthens the gums. Its leaves are beneficial for erysipelas and itching, foul ulcers and skin eruptions, and prevent sweating. Its benefits are many times greater than we have mentioned."

500 years later, Avicenna, one of the greatest Islamist writers and thinkers, included this sound advice in his famous *Canon of Medicine*, which remained the number one medical resource for the Islamic world and Europe through medieval ages.[3,4] Ibn Qayyim al-Jawziyya, who later authored *Medicine of the Prophet*, used the same quote to advise the 13–14th Century Islamic world about the benefits of olive oil[5] (personal communication with Dimitri Gutas).

Today, it may seem that each day brings another scientific report claiming that extra virgin olive oil (EVOO) makes you smarter, healthier, more limber, and longer-lived. At the same time, skeptics maintain that the hype surrounding olive oil as a healthy food is a modern day charlatanism.[6] How do we make sense of the bewildering array of health claims and evaluate for ourselves what and whom we can trust? In this chapter, we will first use what we have learned about the chemistry of the oil to build a general picture based on the known central role of fats in nutrition. We'll present evidence for the understanding of olive oil as a healthy fat and then explore our new understanding of its role in both treating and preventing disease.

8.1 HOW DO OUR BODIES USE FATS AND OILS?

Before getting into specific health claims about olive oil, let's first understand how oils, fats, and their breakdown products are used by our bodies as sources of energy, as building blocks for the formation of cell membranes, and as starting material in the synthesis of hormones and other signaling molecules. We also need to know how different organs in our bodies (like the heart or the brain) have different needs and how serving up a proper meal to each organ often becomes the job of the liver and the pancreas, the organs chiefly responsible for regulation of digestion of fats and carbohydrates.

8.1.1   Getting the Olive Oil to the Cells

Our bodies are mostly water (60% water) and our blood even more so (92% water). Yet, as we explained in Chapter 1, the fats and oils we love are not soluble in water, including our saliva and mostly aqueous digestive system. We need the calories that the energy-rich fats can provide, but how can they be digested? The body uses *enzymes* to break down the intact fats and oils (known as triacylglycerides or TAGs) into more soluble building blocks (the free fatty acids or FFAs) and transport them through the intestines and safely to the liver. There, the liver will assemble fats with special carrier proteins and direct the supply of fats into the bloodstream and out to the rest of the body. Let's take a look at this process.

Toolbox 8.1   What is an enzyme? Take 2.

Enzymes were discussed in Chapter 5 in a discussion of maximizing oil yield during the milling process and developing the fragrances during malaxation. Here, we discuss the more general role of enzymes in human biology. An enzyme is usually a protein molecule that is designed to bind a particular small molecule – choosing it from among the thousands of other molecules in the body – and then direct its chemical transformation into a new molecule. Enzymes are said to catalyze chemical transformations, which means that they will speed up the rates of reactions and make it possible for reactions to happen without high temperatures or high pressures of chemical reactors found in factories or laboratories. The rate at which an enzyme works is adjustable depending on the needs of our body at a particular moment: it can be accelerated, slowed down, or stopped altogether. This is part of the regulation of metabolism.

As shown in Figure 8.1(a), some of the fat is digested in the mouth, where enzymes called lingual lipases begin the process of breaking down TAGs. Swallowing pushes the food through the esophagus and into the stomach. The large TAG fat globules pass relatively untouched through the stomach and into

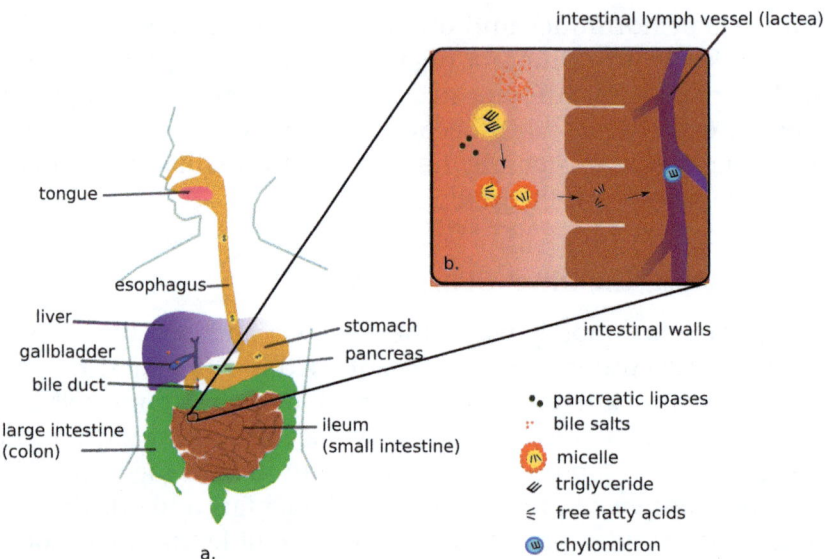

a.

b.

pancreatic lipases
bile salts
micelle
triglyceride
free fatty acids
chylomicron

intestinal lymph vessel (lactea)

tongue
esophagus
liver
gallbladder
bile duct
large intestine (colon)
stomach
pancreas
ileum (small intestine)
intestinal walls

Figure 8.1   The human digestive system including (a) organs responsible for fat digestion and (b) detail of the sequestering of fat, bile salts, and pancreatic lipases into fat globules, the subsequent breakdown into free fatty acids and formation of free fatty acid/bile salt complexes that assemble into micelles, absorption through the cells in the wall of the small intestine, and then reassembly and packaging of triacylglycerides into chylomicrons for transport through the body.

the small intestines, where three key ingredients are mixed. The common bile duct delivers several different lipases (major digestive enzymes for fat), made and secreted by the pancreas, that mix with the TAGs. These enzymes break the fats into FFAs and glycerol. At the same time, cholesterol derived bile salts, made in the liver and stored in the gall bladder, are secreted into the bile duct and pass into the entrance of the small intestines. One job of the bile salts is to enhance the solubility of the fatty acids. These FFA and bile salt pairs will often assemble into collections called micelles, as shown in Figure 8.1(b). These micelles can easily lose or gain FFAs and can therefore pass these FFAs across the intestinal wall. In our intestines, some FFAs may be digested by our gut *microbiome*, but most will eventually pass through the walls of the small intestines and into an intestinal lymph vessel known as a lacteal, where they are once more reassembled into TAGs. Figure 8.1(b) shows how special transport packages called chylomicrons load up a big cargo of TAGs from the lacteal and shuttle them through the bloodstream, delivering some to the peripheral tissues directly for energy and the rest to the liver. Alternatively, FFAs can bind to the carrier protein albumin and be transported through the body.

The liver is the central command post that takes in signals from the rest of the body to decide what to do with the TAG laden chylomicrons. Under certain conditions, the liver repackages the TAG with cholesterol and specific proteins into smaller assemblies called very low density lipoproteins (VLDL) and low density lipoproteins (LDL). These assemblies are released into the blood to deliver the TAG throughout our bodies where it can be used as needed. Cholesterol will be modified and used to adjust the flexibility of cell membranes and as a precursor for many hormones. The TAGs (both delivered to the peripheral tissues and in the liver) can have several different fates, as outlined in the next section.

One important feature of this process is that the FFAs are not altered during this process. This means that the nature of the TAGs circulating through your body is completely determined by what you consume. Consume a diet rich in saturated fats? The TAGs in your bloodstream due to this process will be largely saturated. If your diet contains a lot of olive oil, the major components of the circulating TAGs will be monounsaturated fats.

Toolbox 8.2   What is our microbiome?

Our microbiome is the collective bacteria that call our body home and can be found on our skin and most orifices of our body. The microbiome in our intestinal system is particularly important. Did you know that our gut is host to more than a hundred trillion copies of over 1000 *different* types of bacteria?[7] Whenever we eat, we also feed our microbiome. Sometimes what we eat can actually *change* the types of bacteria as some of them prefer fats or cannot metabolize certain foods. Our overall health is tied to the health of our gut bacteria – they can help us fight infection, process certain foods, and sometimes secrete signaling molecules that adjust some of our own regulatory pathways. New research has been done on the link between our microbiome and obesity.[8] Fortunately for lovers of olive oil, our gut microbiome are happy and healthy when fed oleic acid at the level recommended in the Mediterranean diet.[9]

8.1.2   What Happens to Fuel in Our Cells?

The most important role played by molecular fuels, such as sugars, TAGs and FFAs, is to provide the energy necessary for us to stay alive. Of course we need energy to run a marathon, but we also need energy to breathe, for our hearts to beat, for our nerve cells to fire, and therefore to think and to dream. This energy release must be controlled, literally so we don't burn up. We'll also need to find a way to control location of the energy produced and store it.

Many of the steps involved in the production of energy are carried out not just anywhere in the cell, but inside a special part called the mitochondrion (pronounced "might-o-kon-drion"). Almost a cell within a cell, it has a membrane and a specialized set of enzymes inside for many key cellular functions, including much of the energy production. Discussion of how fats help build membranes such as these will be deferred until Section 8.1.4. For now, it is important to know that all of the energy production from burning fatty acids is generated within the mitochondrion.

8.1.2.1   Burning Food for Energy: Making ATP and NADH.   At the most fundamental chemical level, the energy we need (measured in calories or joules) is derived from combustion reactions, just like the combustion of gasoline is used to power our cars or the burning of wood, coal, or gas is used to heat our homes. All combustion reactions involve the combination of a fuel with oxygen, a process in which electrons belonging the fuel are lost to the electron hungry oxygen. This can happen directly, as in a candle, or indirectly, as in our body. From a chemical perspective, the foods that fuel our bodies all have one thing in common: they contain carbon and hydrogen atoms bonded together. The different fuels (carbohydrates, fats, protein) each have their own special niche. Ultimately for each fuel, the products of combustion are carbon dioxide (CO_2), water (H_2O), and energy that can be used to do work.

Unlike combustion reactions that happen in campfires or in car engines, combustion reactions in our body occur at low temperatures and pressures in a highly regulated step-wise fashion. Our cells contain enzymes to break down carbohydrates, proteins, and fats. There are also some small key molecules that play important roles. First is the molecule adenosine triphosphate (ATP), shown in Figure 8.2, which holds the energy released during combustion and stores it as chemical potential energy.

If this didn't happen, the energy would be released as heat and not available to do work in the body. The energy stored in ATP can be quickly and efficiently released at a later point when it is needed, by enzymatically clipping off one or more of the phosphate groups $(PO_4)^{3-}$. Since ATP exists in every cell of our body and can be at hand whenever and wherever it is needed, it is called the universal energy currency of the cell. Other currencies exist in particular organs or for particular needs, but ATP is accepted everywhere. ATP is used to power muscle contraction, nerve impulses, the synthesis of hormones, and nearly any reaction in the body that requires energy.

A second small molecule that is an intermediate breakdown product in fat, sugar, and protein metabolism is acetyl CoA, first introduced in Chapter 3. Figure 8.3 highlights the acetyl group

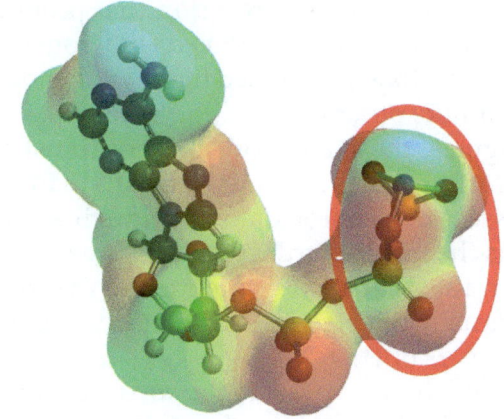

Figure 8.2   ATP – energy stored when the phosphate bond of ATP is made from its reactants and released when the bond is broken.

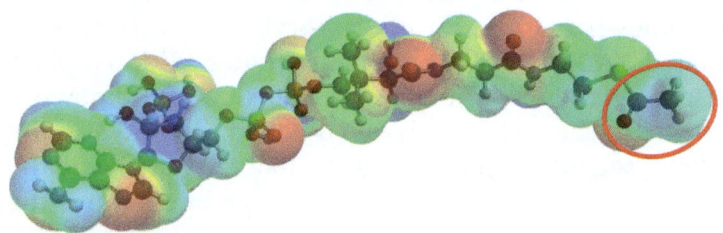

Figure 8.3   Acetyl CoA – carries carbon atoms in the acetyl group, $O=C-CH_3$, shown at the right.

of acetyl CoA that consists of six atoms originating from the original food substance (three hydrogen atoms, two carbon atoms and one oxygen atom or $H_3C-C=O$). The acetyl group is attached through a sulfur atom to the larger molecule, called coenzyme A (CoA). From this point forward in metabolism, a shared set of eight enzymes known as the citric acid cycle is used by all fuels to oxidize the acetyl group to two molecules of CO_2. The waste product CO_2 is disposed of when we exhale.

A third small molecule, nicotinamide adenine dinucleotide (made from niacin, or vitamin B_3), can be found in oxidized (NAD^+) or reduced (NADH) forms. This pair plays an important role in the combustion of food. Oxidation and reduction reactions were previously introduced while discussing antioxidants in Chapters 3 and 4, and you might remember that whenever one molecule is oxidized, another must be reduced. Combustion itself is an oxidation–reduction reaction that produces oxidized CO_2 from our reduced food and reduced H_2O from O_2. As the acetyl CoA made by our food is oxidized in the mitochondria, it loses electrons but, unlike other combustion reactions such as campfires or car engines, these electrons are not transferred directly to O_2. Rather, NAD^+ accepts a pair of electrons to form NADH and the NADH passes the electrons along through a series of small steps known as the electron transport chain, using iron-containing enzymes in the membrane which make the mitochondrion deep red. Eventually the electrons reach O_2, which is reduced to form two molecules of H_2O.

Breaking down the transfer of electrons into many different small steps affords the system a level of control not available when the transfer happens directly. In fact, it is this shuttling of electrons through the electron transport chain that makes it possible for the mitochondria to make ATP through yet another process known as oxidative phosphorylation. NAD^+/NADH (and the related $NADP^+$/NADPH that contain an additional phosphorus atom) together are critical indicators of what is known as the redox state of the cell. This level signals the cell to speed up or slow down sets of coupled reactions like fatty acid oxidation or fatty acid synthesis, balancing the energy budget and thereby affecting the health of a cell and its ability to respond to its environment.

8.1.2.2 Getting the Fuel into the Cells. The first step is absorption of the food from the circulatory system and into the various cells in our body. This absorption will depend on many factors, most importantly the type of cell and the signals that cell has received whether or not to open certain import channels or transport systems.

Complex carbohydrates (polysaccharides) are found in food such as bread and pasta, and contain chains of sugars linked together. They are broken down into simple sugars by amylase enzymes secreted by the salivary glands into the mouth, by the pancreas into the duodenum, and by the lining of the small intestines. Simple sugars such as sucrose (a disaccharide) can be found in foods such as fruit, honey, and jams and can be ultimately broken down into the most simple kinds of sugars, monosaccharides. Glucose, the most useful monosaccharide, can be absorbed by all cells in our body in response to insulin, a hormone that stimulates the glucose transporter in the cell membrane. Sugars, especially glucose, are widely and easily distributed to the cells.

Proteins can also be used for fuel. The breakdown of proteins into their component amino acids is the first step of their digestion and begins with protease enzymes in saliva and in the gut. From the bloodstream, the amino acids are taken up into all cells by specific transport systems. While they can be burned for fuel, their primary function is to be used to build new proteins inside the cell.

Fatty acids are also taken into the cell by transporter proteins.[10] Once inside the cell, they are enzymatically joined to coenzyme A to form a molecule of acyl CoA, just like acetyl CoA but with a carbon chain longer than two carbons. Short chain acyl CoAs can diffuse into the mitochondria – the location for the enzymes that will break them down – by diffusion. Longer chain acyl CoAs have a separate system that enables their transport into the mitochondria.

8.1.2.3 Production of Acetyl CoA, the First Step in the Oxidation of Fat. Now, it's time to look more closely at *metabolism* (derived from the Greek word *metabole*, meaning to change), and specifically how fats are broken down within our body to produce ATP.

Without enzymes, oxidizing a fat in a controlled and systematic way is impossible. In 1904, Francis Knoop made a surprising discovery that fats are broken down (and built up) two atoms at a time. This explains why most fats contain even numbers of carbon atoms. Not only that, the breakdown always starts at the carboxyl end of the fatty acid chain. After one two-carbon unit is removed, the cycle repeats until the fat is completely converted to acetyl CoA.

Knoop realized that this must mean that the bond between the carbon at position 2 (C2) and the carbon at position 3 (C3) is broken. C3 is called the beta carbon because of its location two carbons away from the carboxyl group. As the bond breaks, the molecule loses electrons and is oxidized. Because of this, the entire series of reactions is called "beta-oxidation of fatty acids." As shown in Figure 8.4, the two carbons clipped off bind to CoA to form the now familiar molecule, acetyl CoA. The shortened fatty acid goes through another round of beta oxidation, another two carbons are used to make another molecule of acetyl CoA, and so on until the fatty acid is gone. Thus, a fatty acid such as stearic acid, with its 18 carbon atoms, goes through eight separate cycles of beta oxidation to make nine molecules of acetyl CoA.

8.1.2.4   What Happens to the Acetyl CoA?. The production of several molecules of acetyl CoA from one molecule of fatty acid is by no means the end of the story. At this point, one of several things can occur depending upon the needs of the cell or our bodies. Some of these we have already mentioned, but all are included here for completeness and are summarized in Figure 8.5.

- The most common path is to continue the oxidation process, breaking down acetyl CoA, regenerating the CoA, and eventually producing energy in the form of ATP. This process occurs in the power stations in our cells, the mitochondria. A coordinated series of reactions work to turn the carbons from acetyl CoA into CO_2 (the citric acid cycle), the electrons from the acetyl CoA into NADH to reduce O_2 to water (the electron transport chain). These reactions release energy which is captured as ATP (oxidative phosphorylation).

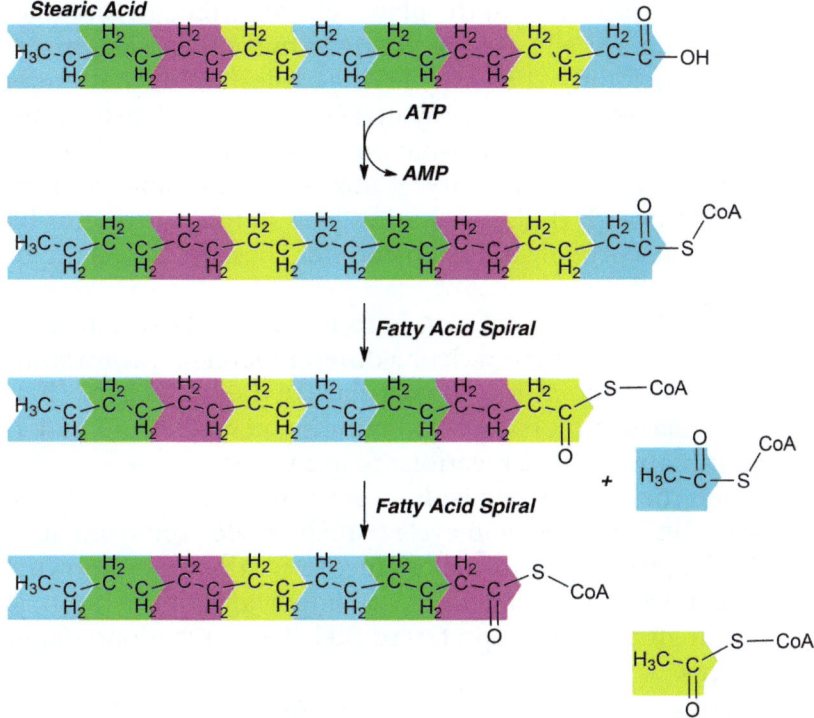

Figure 8.4 Fatty acid breakdown – consecutive bond breakages with each break producing a two carbon unit bound to CoA to form acetyl CoA and a residual fatty acid molecule shortened by two carbon atoms. The bond to be cleaved between the C2-C3 atoms is shown in the image along the edge at which the color changes.

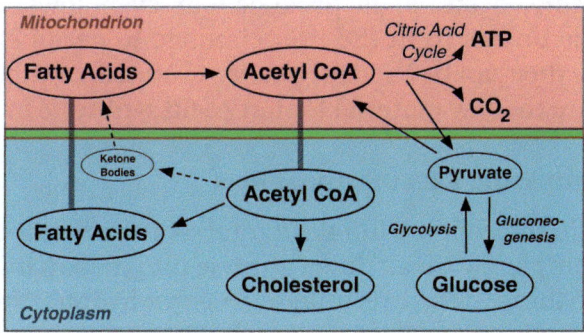

Figure 8.5 Shown are the various fates of acetyl CoA. Which path will be taken depends upon the type of cell and the demands of the body. Reactions in the upper box are carried out in the mitochondria, the bottom in the cytoplasm, each having its own pool of acetyl CoA. The dotted arrows emphasize that, typically, the ketone bodies are only made in the liver and only converted to fatty acids in the brain.

- In a fashion similar to the plant cells described in Chapter 3, some human cells use acetyl CoA to make secondary metabolites, including cholesterol. This cholesterol is the "endogenous" pool that your body makes and not the "exogenous" pool that comes from what you eat. Cholesterol is necessary to adjust the rigidity of our cell membranes and it is made through the same pathway as steroid hormones like estrogen, cortisol, and testosterone. These reactions require energy.
- A third fate for acetyl CoA is to make new fatty acids, perhaps of a different length or a different level of unsaturation. This fatty acid synthesis takes place in the cytoplasm. This is particularly important when building new cells and making membranes. These reactions require energy.
- When the body is particularly starved for glucose, intermediates in the citric acid cycle (which are derived from acetyl CoA) can be used to make glucose in a process known as gluconeogenesis. This might be important for the brain, for which glucose is the preferred fuel. These reactions require energy.
- Finally, through a unique process which happens only in the liver, acetyl CoA can be transformed into a family of small molecules called ketone bodies (simply meaning molecules with a ketone functionality). While the production of ketone bodies is typically low, the two most common ketone body molecules are acetoacetate and β-hydroxybutyrate. A third molecule, acetone, is produced in even lower amounts. Ketone bodies from the liver circulate to peripheral tissues where they can be used for energy, excreted, or exhaled. They also are the raw material for fatty acid synthesis in the brain.

8.1.3   Comparing Fats to Other Fuels

Once inside the cell, it will take a total of 18 different enzymatic steps to completely break down glucose into carbon dioxide. The first 10 of these steps are unique to carbohydrate metabolism and occur in the cytoplasm of the cell. The remaining eight steps constitute the citric acid cycle, the pathway shared by all foods, which takes place in the mitochondria. Each step has its own enzyme. The breakdown of one molecule of glucose produces about 28 molecules of ATP.

Once inside the cell, the metabolism of amino acids is a more complicated process that involves converting the amino acid into an intermediate compound minus the amino group. At this point, enzymes in the citric acid cycle can process the intermediates to harvest the energy. Metabolism of one molecule of the simplest amino acid glycine produces only enough energy to make about eight molecules of ATP. This shouldn't be too surprising, as amino acids are not considered to be a major source of energy, but rather as building blocks for proteins.

There, the oxidation begins and ~118 molecules of ATP are produced from metabolism of one molecule of a fat such as oleic acid. Examination of the energy yield per molecule in Table 8.1 helps to explain why we often refer to fats as "energy dense." While only one example is given here, we should note that there is very little difference between one fatty acid and another in the amount of energy that is released per gram of the acid. The number of food calories in a gram of any fat or oil is very nearly the same.

8.1.3.1   What Does Your Heart Want? Organ Preference for Different Fuels.   It may surprise you to learn that different organs in our bodies have different food preferences.[11] Our brains are very selective, choosing glucose for fuel. The brain needs a constant source of energy – it doesn't have the capacity to store glucose as the storage polysaccharide glycogen, so during starvation it uses emergency rations in the form of ketone bodies that can be served up by the liver. Circulating TAGs and FFAs in the blood are bound up by other proteins and cannot serve as fuels for the brain because they don't cross from the blood into brain tissue. If the brain needs fatty acid to make new membranes, it can use

Table 8.1   Energy density of various foods.

| Class of food | Combustion energy[a] | | Energy density |
	kJ g^{-1}	kcal g^{-1}	ATP molecule/ fuel molecule
Carbohydrates – such as the monosaccharide glucose	17	4	~28
Proteins – such as the amino acid glycine	17	4	~7
Fat – such as the fatty acid oleic acid	39	9	~118

[a]Note: kilojoules (kJ) and kilocalories (kcal) are common units used by scientists. One kcal is equal to one food Calorie (Cal).

the ketone bodies and use them as useful starting materials for the synthesis of longer chain fatty acids.

Skeletal muscles will use nearly any fuel. The most common fuels are glucose, fatty acids, and ketone bodies. Skeletal muscles can store glucose as glycogen and typically have a large supply. It is often the case that, when exercising, the blood cannot keep up with the demand for molecular oxygen, O_2. In these cases, your skeletal muscles have a back up plan: using the first 10 steps of glycolysis, which do not require O_2, can still yield a small amount of energy to keep cells functioning. This only yields a fraction of the energy of the complete breakdown of glucose, but it does provide some ATP and enough NADH to help muscles contract. The muscles can continue to work anaerobically, albeit inefficiently, until the pH drops enough that it becomes too painful. Fatty acid metabolism does not have a comparable anaerobic pathway.

Our heart is also a muscle, but unlike skeletal muscles, our heart must beat constantly without interruption – not on demand as a skeletal muscle. As discussed a bit later in this chapter, our hearts can use glucose but also derive a large amount of their energy from fatty acids. Since most of the enzymes needed to break down these fats are located within the mitochondria of our cells – it won't surprise you to learn that the heart needs a constant supply of oxygen and cannot survive for long without it. The dense concentration of mitochondria in our heart is also responsible for its deep red color. The choice of fatty acids as fuel also allows most efficient transport of energy to the heart. Note that the number of ATP molecules made per molecule of fuel is vastly higher for fatty acids than any other fuel.

8.1.4   Fatty Acids are Also Used to Make Cell Membranes

Fats are also necessary for making our cell membranes, as well as the membranes around the organelles inside our cells (nucleus, vacuole, mitochondria). Membranes are built from phospholipids and have the job of keeping certain materials inside the cell or organelle and other material outside. Proteins in the membrane act like gatekeepers, controlling which materials enter and leave the cell. Figure 8.6 shows the chemical structure of a phospholipid containing a glycerol backbone like TAGs, but with two fatty acids instead of three. A phosphate group replaces the third fatty acid. Since a phosphate group is polar, this modification creates a complex molecule

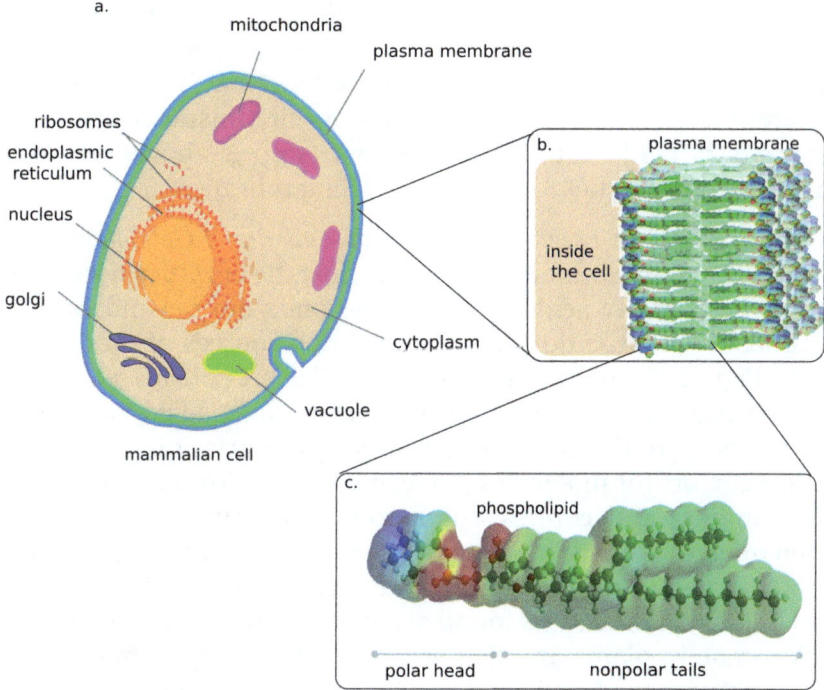

Figure 8.6 Cell membranes are made up of molecules derived from fats known as phospholipids. (a) Shown here is an animal cell with the internal organelles (such as the nucleus, several mitochondria, and vacuoles). Many of these organelles have membranes to separate them from the cytoplasm of the cell. The plasma membrane separates the inside of the cell from the outside. (b) Cartoon representation of a section of the plasma membrane which is made up of a bilayer of phospholipids, nonpolar on the inside and polar on both sides. Note the polar headgroups that face the polar cytoplasm and the polar exterior of the cell. (c) One molecule of a phospholipid showing the polar head group and the nonpolar tail.

in which the head group is hydrophilic and likes to be in water while the nonpolar fatty acid tail is hydrophobic and does not like to be in water. This biphasic aspect of a phospholipid causes them to assemble into micelles, vesicles, or bilayers in which a barrier is created between the inside of a cell and the outside.

The phosphate can be further modified with another chemical group to change its charge and this can affect its function. The length and particularly the unsaturation (number of double bonds) of the fatty acid will make a big difference in the properties of the cell membrane. Too many double bonds and you make the membrane susceptible to chemical oxidation. Too few, and the membrane will become less fluid and the cell can't stretch or transport nutrients like it is supposed to. New fatty acids with different levels of saturation are made as circumstances change. These fatty acids can be synthesized (step 3 above) or used directly from the diet to make new cell membranes.

8.1.4.1   What are Essential Fatty Acids and Can I Get Them from Olive Oil?.   As we humans evolved, we lost the enzymes necessary to synthesize fatty acids with double bonds out beyond carbon 9, probably because our diet provided rich sources of these fatty acids. Figure 8.7 shows two fatty acids which are considered essential and must be present in our diet. These fatty acids, with double bonds beyond carbon atom 9, are very important because they are precursors for molecules that can boost our immune response (eicosanoids) and control our moods (endocannabinoids). We often refer to fatty acids like linoleic acid as an ω-6 ("omega 6") and linolenic acid as ω-3 because, if you begin counting from the left end as shown, the C=C is found at carbons 6 and 3, respectively.

Fortunately, plants and fish never lost the ability to make fatty acids with double bonds at the ω-6 and ω-3 positions. Because of this, plant and fish oils are good sources of both linoleic and linolenic fatty acids. While olive oil is primarily composed of monounsaturated oleic acid, it does have smaller amounts of a dozen other fatty acids including the polyunsaturated ω-6 linoleic acid (about 10%) and the ω-3 linolenic acid (about 1–2%).[12] Consumption of 60 mL olive oil per day provides about 6 g of ω-6 linoleic acid, thereby satisfying the recommended daily consumption of 5–6 g of this essential fatty acid.[13] Since the recommended daily consumption for ω-3 fatty acids is 1.0–1.5 g and the olive oil consumed at the

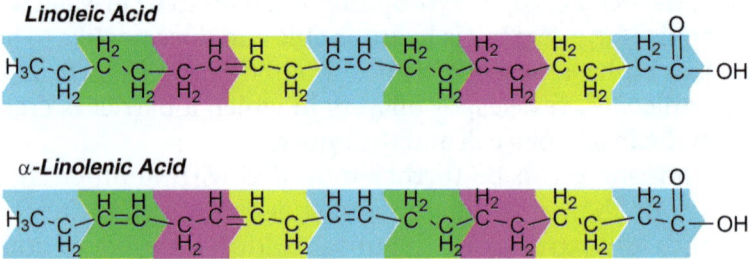

Figure 8.7   Essential fatty acids ω-6 linoleic and ω-3 linolenic acids.

recommended level provides only half that amount, other sources such as flaxseed oil, chia seeds, walnuts, fresh basil, and canola (rapeseed) oil can supplement (Linda Costa, personal communication). Certain fatty fish such as salmon, mackerel, sardines, and tuna are a rich source of the essential ω-3 fatty acids, and the American Heart Association (AHA) recommendation is that adults consume a 3.5 ounce portion of fish at least twice per week. This might be the time to point out the wonderful food pairing of fish and olive oil that is another staple of the Mediterranean region.

8.1.5   Hungry? Full? What, When, How, and Why of Appetite and Its Regulation

Appetite and fullness (satiety) are two incredibly important concepts that have a biochemical basis. At the simplest level, our appetite is created by two hormones: ghrelin and leptin.[14] Ghrelin is made in cells in the digestive system to act as a monitor of the state of the stomach. When your stomach is empty, the ghrelin is first activated by the attachment of an eight carbon long fatty acid to the hormone and then secreted into the bloodstream where it is quick to act at two sites. It can travel through the bloodstream up to the brain, where it binds to a receptor in the hypothalamus and sends a signal to your brain, "I'm hungry! Eat!" Another set of receptors in your stomach receives the signal and initiates borborygmi, the audible gurgling and rumbling of an empty stomach.

Toolbox 8.3   What is a hormone?

A hormone is a molecule that is used in our body to carry instructions from one part of the body to another. A hormone can be made up of large or tiny protein molecules (known as peptide hormones), steroid based chemicals such as estrogen or testosterone, or other types of molecules. The chemical signal can travel very short distances from one cell to its neighbor (the endocrine system) or very large distances through the blood or lymph (the exocrine system) and transmit information that might cause a nerve cell to fire in your brain, a heart cell to switch from one type of metabolism to another, a glucose transporter in a cell membrane to open its gates to allow glucose to enter a cell, or a muscle cell to contract.

Once you have eaten enough to fill your stomach, the slower acting hormone, leptin is secreted from adipose tissue (fat storage cells), through the blood and up to the hypothalamus to signal "I'm satisfied! Stop eating!" to your brain. Leptin also steps up the metabolic rate of other cells to make sure to burn the food that has been eaten. Leptin is a most important enzyme in weight control. Studies have found that the leptin receptors in many people have become insensitive to leptin – a condition associated with obesity. No matter how much is eaten, the "satisfied" signal is never received by the brain. Not surprisingly, leptin is a potential therapeutic target for those individuals who are chronically obese.[15]

Have you ever had a strong desire for a particular food, whether it is a sugary donut, a protein rich steak, or a pint of ice cream? While there are certainly cultural and psychological components that drive us toward or away from certain foods, the recent discovery of taste-like receptors in our digestive system and elsewhere in our bodies shows how our appetites for particular foods are shaped.[16] Receptors have been found in the stomach for bitter compounds such as those found in EVOO. These receptors can cause the release of ghrelin, stimulating our appetite. Perhaps this is why so many "appetizers" and "aperitifs" contain bitter components. The key here is that the appetite is stimulated when small amounts of the bitter components are eaten. Over-stimulation of the stomach's bitter receptors for longer periods of time actually ends up inhibiting ghrelin activation and suppressing appetite.[17]

While this book is about olive oil, we can't ignore the inter-relatedness of all the foods that we eat. Most people are more familiar with the hormone insulin, which is secreted by the pancreas in anticipation of a good meal and activates glucose transporter proteins in membranes to load sugars from the bloodstream into target cells. If the body has cleared the glucose from the blood, either between meals or after a period of intense exercise, another hormone, glucagon, is secreted from the pancreas that tells the body that glucose is gone, and the stored polysaccharides (glycogen) in the body need to be broken down for energy. Our cells don't have large reservoirs of glycogen, and so when those reserves are gone, it is the adipose tissue, the fat cells, that can begin the breakdown of stored fat for the energy needs of our body.

8.1.6   Regulation of Metabolism

As in any complex system, regulation is extremely important. What food we choose to give our body dictates the body's response. When we eat fats and oils, an assessment is made of the total lipid load in the blood. This lipid load includes FFAs bound to proteins, TAGs with VLDL and chylomicrons, and acetoacetate and other ketone bodies. If we need energy and our lipid load is high, hormone sensitive lipases are released, breaking down more TAGs to allow FFAs to be metabolized in the mitochondria. In the cytoplasm, a substance known as C75 shuts down fatty acid synthesis.[18] If we don't need energy and our lipid load is high, the body can block metabolism of FFAs by blocking their entrance into the mitochondria, and instead, synthesize new TAGs for storage in adipose tissue (fat cells). If the cell requires new fatty acids for the synthesis of membranes, or if certain steroid based hormones need to be made, then fatty acid synthesis in the cytoplasm will be activated and the oxidation inhibited. Figure 8.8 outlines some of the major control points in fatty acid metabolism.

Similar regulation happens for carbohydrates and proteins. A needs assessment and an inventory is done – certain pathways are stimulated while others are inhibited. When our carbohydrate load is high, insulin is released which will stimulate formation of glycogen and fat, deactivate hormone sensitive lipases, and activate the acetyl CoA carboxylase for fatty acid biosynthesis and packaging to VLDL for storage in fat tissue. Couple this biochemical fuel assessment with the appetite stimulating (ghrelin) and fullness (leptin) hormones, and we have a well-oiled (so to speak) machine that has served us well throughout human existence.

8.2 HOW REAL ARE THE HEALTH BENEFITS OF EXTRA VIRGIN OLIVE OIL (EVOO)?

Perhaps the most frequently held wish for a loved one – a toast made at a birthday or anniversary – is "Here's to a long and healthy life!" We want these longer and healthier lives for ourselves and our loved ones to be ones in which we are alert, mobile, and continue to have meaningful relationships with our friends and family. When we study the health benefits of EVOO, we will have to focus on those benefits that we can measure, typically those that are quantifiable. In this chapter, we will discuss three types of investigations that are used by researchers to make their various claims about the health benefits of olive oil: epidemiological studies, lab based research (*in vivo* studies), and biochemical experiments at the molecular level (*in vitro* studies).

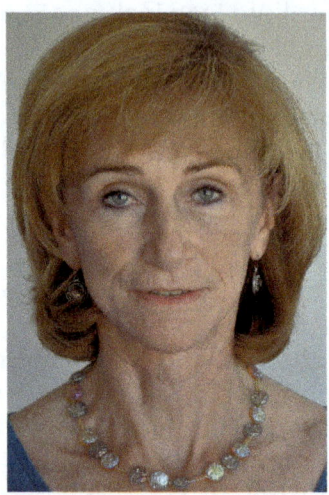

Cameo 8.1   Linda Costa, Stellenbosch, South Africa.

The daughter of Nino Costa and granddaughter of Ferdinando Costa (one of the founders of the olive industry in South Africa), Linda Costa is knowledgeable, outgoing, and infinitely knowledgeable about the olive growing culture in South Africa. Linda's educational background in chemistry, physiology, pharmacology, and microbiology makes her an invaluable consultant to olive enthusiasts in the region. She has written a quick reference primer on the health benefits of olive oil for *The Guide To Extra Virgin Olive Oil*. Together with a friend, Sandra van Schaik, she started a business, "Olive Go Wild" that markets olive oil in novel Vacu-Fresh packaging. The vacuum packaging protects the oil from oxidation because, once the oil is opened, air is never introduced into the remaining oil, as would typically happen after you open a bottle of oil. Drawing on experiments in her own farm, she has also prepared an e-book, *Table Olive Processing – Made Easy*, on the curing of table olives. We confess that one taste of Linda's "Simply Delicious" dry cured Kalamata olives made us converts for life. She runs olive tasting workshops and is often asked to act as a judge in olive oil competitions. In 2010, she was awarded lifetime membership of the South African Olive Industry Association. Linda has her own olive farm, "Awakening," outside of Stellenbosch and uses this as a base for her writing, olive curing, and coordination of her many business ventures.

Epidemiological studies (the analysis of the prevalence of certain diseases in particular human populations) should be done with large numbers of people to ensure that results are both significant and representative of the population. The study design can include an intervention, such as asking some of the participants to do something they would not normally do. Participants might take a pill or drink a liquid or exercise for 20 minutes every day,

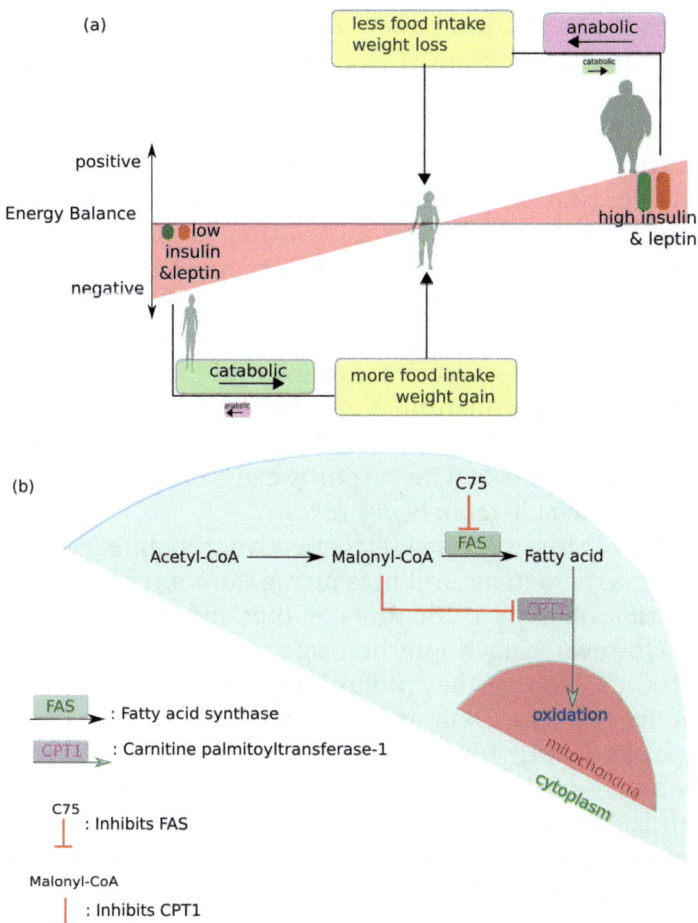

Figure 8.8 Regulation of metabolism (a) as a balance of catabolism and anabolism, as determined by an energy balance. If the energy balance is low, hormones insulin and leptin are low, catabolic breakdown processes are shut down, and anabolic processes are stimulated to cause food intake. If energy balance is positive, the hormones' levels increase, catabolism is stimulated, and anabolic processes are shut down; (b) at the cellular level, malonyl CoA stimulates transfer of fatty acids into the mitochondria for oxidation while the inhibitor C75 can slow down synthesis of new fatty acid. Adapted from Macmillan Publishers Ltd: *Nature Reviews Neuroscience*, Seeley and Woods, **4**(11), copyright 2003.

and are compared to other participants who receive no intervention (often referred to as a "control group"). You may hear
of studies that are done under "blind" conditions in which the
participants do not know if they are in an intervention group
or a control group, or "double blind" in which the researchers themselves are also not told which participants are in the
intervention group and which are not. These studies must be
prospective; that is, enrolling a host of participants and then
following them forward in time. Another prospective approach
is to rely on the natural variation in practices among the members of a large group and does not include any intervention
(non-intervention) – so that a health outcome is measured
based on an existing practice. For example, perhaps some participants naturally consume a lot of olive oil and others hardly
any. The advantage of a non-intervention prospective trial is
that the participants can be carefully examined before the study
begins, and the self-reports are recent.

It is possible to carry out a retrospective trial: interviewing people about past practices and measuring outcomes in the present.
This carries obvious difficulties if diet information is self-re-
ported. However, they might be useful to use as a tool to design
other studies. Most of the epidemiological studies cited here are
prospective studies. Data are collected at the start about each
participant's health history, starting with a full medical exam.
Questionnaires ask them regularly what and how much they eat
and drink, how much they exercise, and such information as
whether they smoke or use drugs. Physical exams and hospital
records form a part of the ongoing study. Over a period of time,
usually for years, this group (called a cohort) is followed. When
the study is about the health benefits of olive oil, the researchers pay particular attention to the consumption of olive oil. The
group can often be separated into those who eat more or less
olive oil, or there can be a control group with a standard diet and
an intervention group who eat increasing amounts of olive oil as
part of their daily diet. In many parts of the world, this variation
can be found naturally within a population. As trends become
apparent, the results are published and results are reported as
relative risks for diseases.

Often these studies report contradictory results. How is this
possible? When evaluating for yourself whether or not to take

the study seriously, you should ask a few questions. In general, a study will be more useful to you if the number of participants is larger, the time over which the study was conducted is longer, the integrity (and independence) of the group who did the study is higher, and the cohort of people studied is close to your own background. Of course, you should always consult your own primary care physician.

The following studies are considered to be the most comprehensive and have had the greatest impact by convincing doctors and other health care professionals to recommend to their patients that they modify their diets to include (among other things) more olive oil:[19]

- *PREDIMED* (Prevención con Dieta Meditteránea): an intervention study of 7447 volunteers between 55 and 80 years old in Spain, focusing on cardiovascular disease (CVD), that has run for five years.[20]
- *SUN* (Seguimiento Universidad de Navarra): the SUN project is an ongoing prospective Spanish study with more than 20 000 middle-aged university graduate volunteers who report every two years. Early results revealed that olive oil consumption was associated with lower blood pressure.[21] Recent analysis shows adherence to a Mediterranean Diet is positively correlated with overall reduced mortality.[22]
- *Three City Study*: a prospective study of the French cities of Dijon, Montpellier, and Bordeaux. From 1999 to 2012, the relationship between vascular disease and dementia was investigated in a population of 9294 men and women over 65 years old. In particular, the relationship between consumption of olive oil and stroke was measured over five years.[23]
- *EPICOR*: an Italian prospective study of ~30 000 women between the ages of 35–74 who live in the northern cities of Turin and Varese, central city of Florence, and the southern cities of Naples and Ragusa. EPICOR has been going on for more than eight years and focuses on CVD.[24]
- *EPIC-Spain*: a Spanish prospective study of over 40 000 participants (38% male) from the regions of Asturias, Granada, San Sebastian, Murcia, and Navarra. EPIC-Spain has been ongoing for over 11 years with a focus on CVD.[25]

Why isn't a study based in the United States in this most elite group? In fact, *MrFIT* (Multiple Risk Factor Intervention Trial) was an important US study that began in 1972 with 12 866 participants who were all at high risk for CVD. It examined the effect of diet and exercise on coronary heart disease. However, in this study, the intervention focused on diets and drugs to lower cholesterol and lower blood pressure and behavioral intervention to stop smoking. After seven years, the defined endpoints for men in the special intervention group did not vary significantly from those who did not receive any intervention, and so the study was brought to a close.[26,27] More recent re-analysis of the data using slightly different endpoints has supported lifestyle changes that were found significant in these other trials: exercise, diet, and adjuvant pharmacological therapies.[28]

Hundreds of epidemiological studies have been done on the health benefits of olive oil in the last few decades. Often, the data and results for different studies are compared one to another in what is called a "meta-analysis." The hope is that by searching for trends or assaying risk across many studies, study biases or methodological weaknesses can be eliminated and results common to all studies might therefore be held to be more universally true.

As these epidemiological studies are being carried out, scientists begin to discover correlations between human behavior and health outcomes. Unfortunately, correlation merely tells us that two measurements behave in related ways. It might be a real relationship (exercise and cardiovascular health), it might be accidental or based on a third related variable (income and cardiovascular health). Deciding how relevant the correlation is requires more testing. This is where lab based studies using animal models or done on isolated non-living systems can play a key role. To verify the relationship, there should be a physiological or biochemical reason for the relationship. Sometimes called a "mechanism," it is a very important part of validating the epidemiology.

8.2.1   (At Least) Three Molecules in Extra Virgin Olive Oil (EVOO) that Keep Us Healthy

One of the best ways to prevent disease is to stay healthy. Maintaining a healthy weight, sleeping well, exercising, having meaningful relationships with others, and maybe a pet or two are all behaviors associated with long life. What are the chemical compounds found in EVOO that help us achieve that goal of staying healthy? We will highlight three here. In the following section, we will discuss our evidence from population studies and experiments in the lab.

8.2.1.1 Oleic Acid – A Gold Standard for Fats. As we now know, the major ingredient of olive oil is TAGs or fat. Within those TAGs, oleic acid contributes approximately 70% of the fatty acid chains in *all* olive oils (including refined olive oil). With just one double bond, oleic acid is a monounsaturated fatty acid (MUFA). The bend in the molecule created by the double bond keeps the molecules from packing closely together and so olive oil is a liquid at room temperature. By contrast, saturated fatty acids (SFAs) have no double bonds, pack tightly against one another, and are solids at room temperature (think lard and bacon grease). Polyunsaturated fatty acids (PUFAs) have two or more double bonds, and so are also liquid. All of these fats and oils are energy dense, so can provide plenty of ATP for our bodies to use. The relevant question here is whether or not oleic acid makes the most healthy fat.

Let's review the different kinds of fats that can be found in food with an eye to the health benefits. What about SFAs? Dozens of studies over several decades have shown a correlation between consumption of high levels of SFA and high serum levels of LDL-cholesterol. These in turn are associated with clogged arteries and heart disease. In both *in vivo* and *in vitro* studies, SFAs have been shown to promote inflammation and enhance the formation of blood clots and plaques on the lining of the arteries.[29] As a result, the American Heart Association suggests limiting our consumption of SFA to less than 10% of our total calories. Two recent studies challenged these results and claimed that their own analysis of the data from multiple epidemiological studies failed to show that reducing dietary SFA was associated with a lowered risk of heart disease.[30,31] These controversial findings got a lot of attention and headlines in the popular press. However, the response of the medical community was swift to point out the limitations of these reports. A 2016 review makes the following useful summary:[29]

"...Among other methodological issues, the authors of the meta-analysis failed to consider the impact of the replacement nutrient, which is critical in analyses of a nutrient that makes up a substantial proportion of total energy intake. In other words, they did not ask, "compared to what?" It is not enough to analyze saturated fat intake without considering a comparator, which is typically refined carbohydrates and added sugars. The lack of association between saturated fat and coronary heart disease in observational studies does not mean saturated fat is benign; it simply means that high saturated fat diets and high refined carbohydrate diets are equally detrimental to heart health."

PUFAs and MUFAs are generally considered to be healthy fats. These fats are derived from plant oils such as olive, avocado, or vegetable oil and seed and nut oils such as canola, peanut, hazelnut, or walnut oil. Each source produces a different fatty acid profile and different ratios of PUFA/MUFA/SFA. Vegetable oils such as corn oil, soybean oil, and sunflower oil usually have the highest fraction of PUFA. Many fish contain oils, such as those found in salmon, mackerel, and anchovies, that are rich in the PUFA ω3 and ω3–ω6 fatty acids that are essential to our diets. One risk factor associated with PUFAs is their tendency to be oxidized by reactive oxygen species (ROS). As detailed in Chapter 6, the rate of oxidation is substantially higher in PUFA than in MUFA. As a MUFA, oleic acid will be much slower to oxidize. For all these reasons, oleic acid can be considered to be extremely healthy.

Trans-fats are not found naturally. They are created in small amounts by partially hydrogenating vegetable oil in an attempt to create a more solid margarine-like product that has a longer shelf life. In this process, PUFAs are converted both to SFAs and MUFAs. Some of the MUFAs can have the unnatural *trans* configuration. *Trans* fats can be found in many grocery items, from cookie or biscuit dough to popcorn and snack foods. Food labels that read "0 mg" of *trans* fats are misleading since food manufacturers are allowed to round down. That means that 0.49 mg of *trans* fat will read as 0 mg. Checking to see if "partially hydrogenated vegetable oil" appears on the ingredient list will give you a "heads up" and allow you to avoid, or at least minimize, your family's consumption of *trans* fats. Another source of *trans* fat comes from poorly tended deep fat fryers. Because the *trans* fat is more stable than the natural *cis* form, oil changed after too much use can accumulate the *trans* form. Research has shown that *trans* fats both raise "bad" LDL and lower "good" high density lipoprotein (HDL), thereby raising the risk of CVD. In June of 2015, the US Food and Drug Administration characterized *trans* fats as "not generally recognized as safe" and has ordered that manufacturers remove them from all foods in the next three years.

8.2.1.2 Antioxidant Polyphenols. The second most important family of molecules in EVOOs work to fight the destructive effects that oxygen can have in our bodies. Our relationship with oxygen can certainly be characterized as a love–hate relationship. We love it because it is essential for the aerobic oxidation that we need to extract energy from our food. How can you burn anything without oxygen? We hate it because it can be quite damaging when it gets where it is not supposed to be and starts wreaking havoc with our proteins, DNA, and membranes. Many human diseases and the cumulative effects of aging itself are a result of oxidation due to ROS such as oxygen, peroxides, and oxygen radicals. The second health benefit of EVOO will be the *presence of antioxidants* in the form of *phenols and polyphenols*. These compounds come from the fruit when it is pressed but are lost if the oil is refined. The particular cultivar, time of picking, terroir, processing conditions, storage conditions, and even how it was shipped to you and stored in your supermarket ALL will affect the amount of phenolics left in the bottle. Phenolics will taste bitter, so a lovely smooth olive oil with no bitterness will provide the MUFA mentioned above but will be bereft of the power of antioxidants. Since, as of this writing, producers are not required to put the phenolic content on the label, how can the health conscious consumer know how to choose the most healthy oil? Fortunately our taste buds are finely attuned to these bitter compounds, so find yourself an EVOO whose taste you enjoy and it will surely contain healthy antioxidants. The power of antioxidants to keep you healthy will be discussed in some key epidemiological studies as well as some animal and molecular studies.

8.2.1.3 Anti-Inflammatory Oleocanthal. Finally, at least one particular compound in olive oil, *oleocanthal* has the power to act as an *anti-inflammatory molecule*. New research has indicated it is active not only against arthritis and stroke, but on cancer and dementia. Oleocanthal's multiple therapeutic effects are beginning to be understood. While no olive oil labels report the quantities of oleocanthal, you will know if it is present by a pepperiness that you can feel in the back of your throat that can cause you to cough. This "pungency" is a telltale sign of the presence of oleocanthal and, in fact, the same sensation one gets from the medicine ibuprofen. We will review some exciting new work that is being done to understand how this molecule is working at a biochemical level.

8.2.1.4   Micronutrients Cannot Be Ignored.   Our bodies require daily consumption of relatively large amounts (gram quantities) of the basic nutrients: carbohydrates, sugars, fats, and proteins. These are burned to give energy and used as raw materials to build our bodies. By contrast, micronutrients are chemicals that must be present in small quantities to assure the proper functioning, growth, and regulation of an organism. Many vegetable and seed oils are processed using high temperature extraction or solvent refinement through which they are stripped of many of these health giving micronutrients that may have been present in the original fruit, seed, or nut. EVOO is never processed this way, and so its natural goodness is preserved. With each tablespoon of EVOO you will also be providing your body not only with fats, antioxidant polyphenols, and anti-inflammatory oleocanthal, but also with small but significant amounts of several micronutrients. There will be α-tocopherol (the most biologically active form of vitamin E, itself a powerful antioxidant and regulator of cellular activity) and squalene (a triterpene that is important in the synthesis of sterols and vitamin D). These compounds are yet another way that you stay healthy and fight disease best with EVOO but will not be a focus of the later sections.

8.2.2   Nutrition's Role in Prevention of Cardiovascular Disease (CVD)

In 1999, a study in Lyon, France, reported a 75% reduction in heart attacks (one cause of which is shown in Figure 8.9) in a group that followed a Mediterranean Diet.[32] Shortly after this, the AHA published an "Action Alert" to bring these results to the American public and suggest that those who are at risk for CVD consider following a similar diet.[33] Since then, other studies in

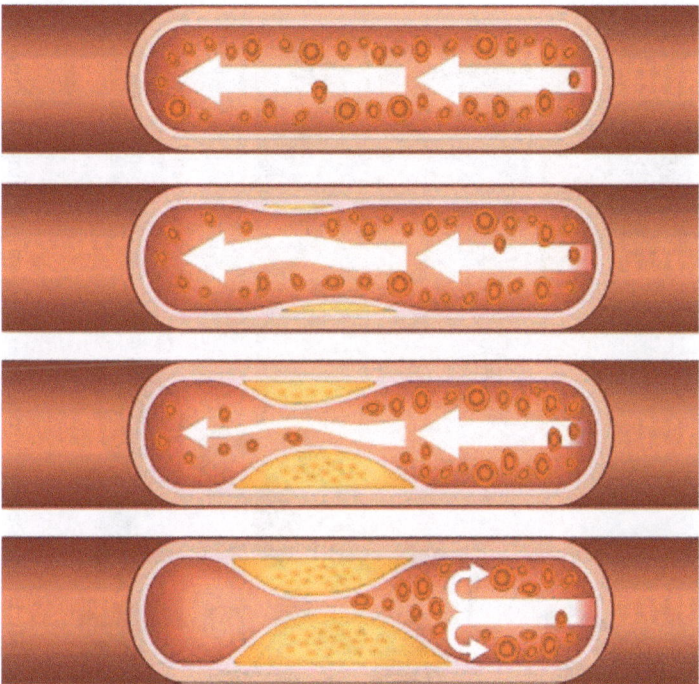

Figure 8.9   Cardiovascular disease (CVD) is the number one killer of men and women worldwide – each year CVD is responsible for 25% of deaths in the US and 30% of deaths worldwide. CVD is characterized by compromised heart function that can lead to heart valve problems, arrhythmia, heart attack, and strokes. CVD is caused by atherosclerosis, or a build up of plaque in the walls of arteries that narrows them and makes it possible for damaging blood clots to form.

Spain (PREDIMED) were commissioned. In 2014, the *Journal of the American Heart Association* published a review of the studies from 1957 to 2013.[34] A whopping 30% reduction in risk of developing CVD can be achieved by following a Mediterranean Diet (PREDIMED) and a 40% reduction in CVD mortality from the highest quintile of consumers (SUN). The American Heart Association, with even more confidence, recommended those individuals at risk of CVD or other related diseases (stroke and diabetes) should follow a Mediterranean Diet.[26]

A Mediterranean Diet food pyramid featuring olive oil as a key daily nutrient that provides 30% of your calories is shown in Figure 8.10. Daily consumption of four tablespoons (about 60 mL)

Figure 8.10 Mediterranean Diet food pyramid (Anna Bach-Faig, Elliot M. Berry, Denis Lairon, Joan Reguant, Antonia Trichopoulou, Sandro Dernini, F. Xavier Medina, Maurizio Battino, Rekia Belahsen, Gemma Miranda and Lluís Serra-Majem, Mediterranean diet pyramid today. Science and cultural updates. *Public Health Nutrition*, 14(12A), 2274–2284, Figure 2 reproduced with permission from Cambridge University Press).

Sweets ≤ 2s

Red meat < 2s
Processed meat ≤ 1s

Eggs 2-4s
Legumes ≥ 2s

Herbs / Spices / Garlic / Onions
(less added salt)
Variety of flavours

Olive Oil
Bread / Pasta / Rice / Couscous /
Other cereals 1-2s
(preferably whole grain)

Water and herbal
infusions

Biodiversity and seasonality
Traditional, local
and eco-friendly products
Culinary activities

s = Serving

Weekly

Potatoes ≤ 3s

White meat 2s
Fish/Seafood ≥ 2s

Every day

Dairy 2s
(preferably low fat)

Olives / Nuts / Seeds 1-2s

Every Main
Meal

Fruits 1-2 | Vegetables ≥ 2s
Variety of colours / textures
(Cooked / Raw)

Regular physical activity
Adequate rest
Conviviality

2010 edition

of EVOO provides you with 480 calories of dense packed energy. The most recent nutritional recommendations from the US Department of Agriculture included following a diet that focuses on variety, nutrient density and amount and limits calories from added sugars and saturated fats. In particular, it suggests including oils and limiting saturated fats to less than 10% of calorie intake per day.[35]

8.2.3 Exactly How Does Olive Oil Help with Cardiovascular Disease (CVD)?

The predominant fatty acid in olive oil is oleic acid. Your heart loves oleic acid and so should you. Take a minute to take your pulse – do you feel it? When doctors take an EKG or listen through a stethoscope, they find that your heartbeat is actually a rhythmic double beat first caused by an atrial contraction that pushes oxygenated blood from the atrium (entryway) into the ventrical, and second a ventricular contraction as that blood is pushed out either to your lungs or to the rest of your body. That steady two beat muscle contraction of your heart starts soon after conception and accompanies us throughout our lifetime. A constant and reliable source of energy is critical and an energy dense source preferable. Studies show that, given a choice, the heart will metabolize fats over any other nutrient.[36] Not only that, if injured hearts[37] or heart cells[38] are treated with oleic acid, recovery is faster and more complete (even so, if you have a heart attack, first reach for the phone, not a bottle of olive oil).

The preponderance of evidence should convince you that a Mediterranean Diet reduces your risk for CVD. But, to many, there remains the question of exactly how this diet works on the human cardiovascular system. A new review published in the *American Journal of Medicine* outlines a multitude of ways that food can be cardioprotective, as shown in Figure 8.11.[39] Key players here are TAG, cholesterol, and the two major carrying proteins LDL and HDL. We have seen that TAG is packaged into the LDL along with cholesterol at the liver for dissemination to the rest of the body. HDL on the other hand is made by cells and functions to scour the body for cholesterol, returning it to the liver for excretion or repurposing.

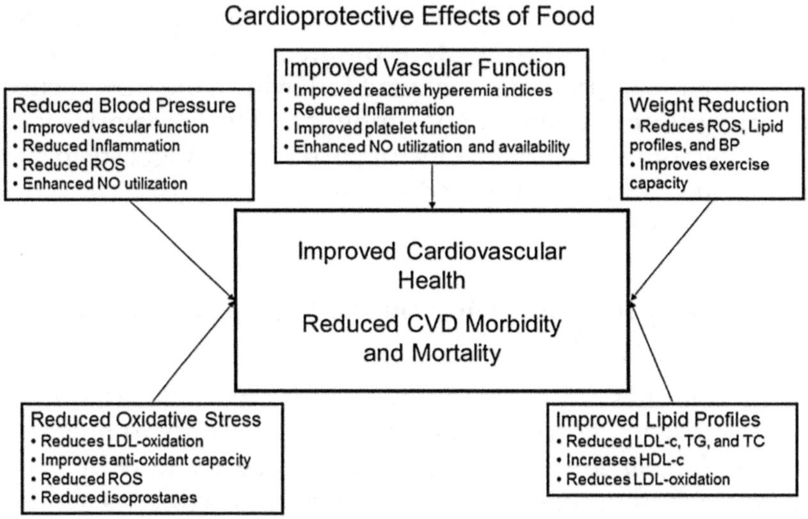

Cardioprotective Effects of Food

Figure 8.11 Ways in which food can be heart healthy (Reprinted from R. Jay Widmer, Andreas J. Flammer, Lilach O. Lerman and Amir Lerman, The Mediterranean Diet, its Components, and Cardiovascular Disease, *The American Journal of Medicine*, 128(3), 229–238, Copyright 2015, with permission from Elsevier).

One benefit of a cardioprotective food is improved lipid profiles in the blood. This means lowering the LDL-cholesterol to below 130 mg dL^{-1}, raising the HDL-cholesterol above 40 mg dL^{-1}, and holding the total cholesterol below 200 mg dL^{-1}. Some doctors prefer calculating a ratio of the total cholesterol to the HDL-cholesterol. The Mayo clinic suggests that a ratio of less than 3.5 to 1 is optimal.[40] As the ratio increases, the risk of developing a heart condition rises. The increased ratio suggests too much LDL – and this is a concern since the lipids can become oxidized and deposit as a fatty sediment in the vascular system, eventually forming plaques.

EVOO is a cardioprotective food. When used in moderation, it improves vascular function by reducing inflammation,[41] reducing ROS, and improving antioxidant capacity. Roberto Carnevale and colleagues at the University of Rome have begun to unravel exactly how this works by focusing on "post-prandial oxidative stress" or the rise in ROS after we have eaten a meal.[42,43] In the first study, published in 2014, 25 healthy individuals were randomly assigned to eat a meal containing 10 g EVOO or not. Several biochemical markers of oxidative stress were measured in the participants' platelets and serum before and two hours after the meal. Two months later the same group was asked to consume a meal with EVOO or corn oil. In both cases, EVOO triggered several protective responses, lowering the production of ROS by blocking NOX2, one of the key enzymes responsible for the production of superoxide, a potent ROS. These *in vivo* studies were followed up by studies *in vitro* that confirmed the findings. A second study, published in 2016 by the same research group, analyzed several other biochemical markers of post-prandial oxidative stress in 30 pre-diabetic individuals and showed once again that the EVOO was protective in that it increased serum levels of insulin, lowered the levels of blood glucose associated with the post-prandial oxidative stress, and lowered serum triglyceride levels. Serum levels of cholesterol and HDL-cholesterol did not differ significantly from controls. While olive oil's ability to raise the HDL-cholesterol is not clear, other evidence shows that the LDL-cholesterol is lowered, and therefore the ratio of total cholesterol/HDL-cholesterol is lowered, which is what really matters. Blood pressure is reduced as a consequence of the improved vascular system and the reduced inflammation. Olive oil enhances the signal carried by nitric oxide (chemical formula NO), a molecule whose presence helps lower and maintain proper blood pressure.[39]

8.2.4 Concerns About Olive Oil and Body Mass

Despite the concerns that so many health and body conscious Americans have about eating fats, eating olive oil in the manner consistent with the Mediterranean Diet, *i.e.* in moderation and along with healthy amounts of fruit, vegetables, and fish, will not make you fat. Individuals participating in the SUN study were placed into five groups (quintiles), each with a different amount of total olive oil in their diet. The results of the odds ratio (chances of gaining weight compared with that of a control group) for these five groups over five years are shown in Figure 8.12. Consumption of the highest level of olive oil by individuals in the top quintile did not show significantly increased body mass. This result – the lack of weight gain in populations with the highest olive oil consumption – was confirmed in two recent studies in different populations.[44,45] This paradox might be

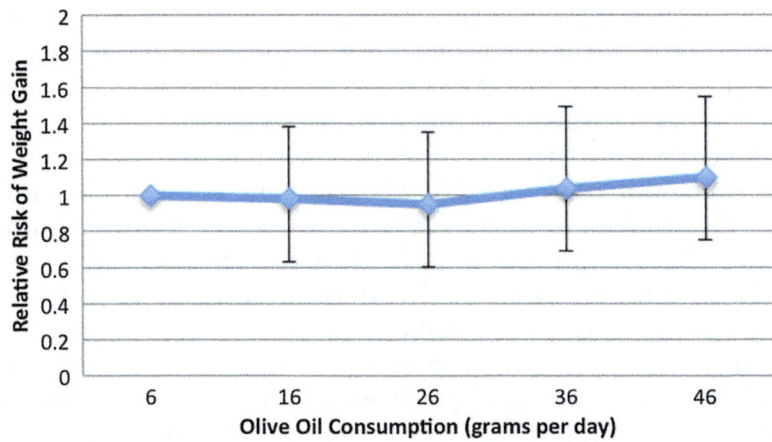

Figure 8.12   Increased olive oil consumption does not lead to weight gain, as seen in a study of 5356 participants in the SUN Study (adapted from Maira Bes-Rastrollo, Mario J. Soares, Miguel A. Martinez-Gonzalez, Maira Bes-Rastrollo, Mario J. Soares and Miguel A. Martinez-Gonzalez, *Olives and Olive Oil in Health and Disease Prevention*, Chapter 96, Copyright 2010, with permission from Elsevier).

explained by the fact that those individuals have increased production of body heat (thermogenesis), increased rate of fatty acid oxidation, or increased excretion of fat through fecal loss.

Understanding why this is true requires understanding the different metabolic fates of different fatty acids. A recent study found that monounsaturated fats such as oleic acid are preferentially taken up and transported by carrier proteins. This makes them more accessible to the metabolic enzymes responsible for their digestion and therefore they are preferentially oxidized to give us energy. Because of this, consumption of foods such as olive oil that contain high oleic acid results in a lowered risk to obesity compared with consumption of other fats.[46] The bottom line: don't feel guilty about dressing your salad or grilled fish with some fine EVOO. It will keep you healthy, taste good, and satisfy you all at the same time.

8.2.5   Phenolics: Antioxidants that Fight Oxidative Stress

Biologically active micronutrients such as phenols and polyphenols act as powerful antioxidants that can operate in a number of ways, as shown in Figure 8.13, to neutralize ROS and to enable

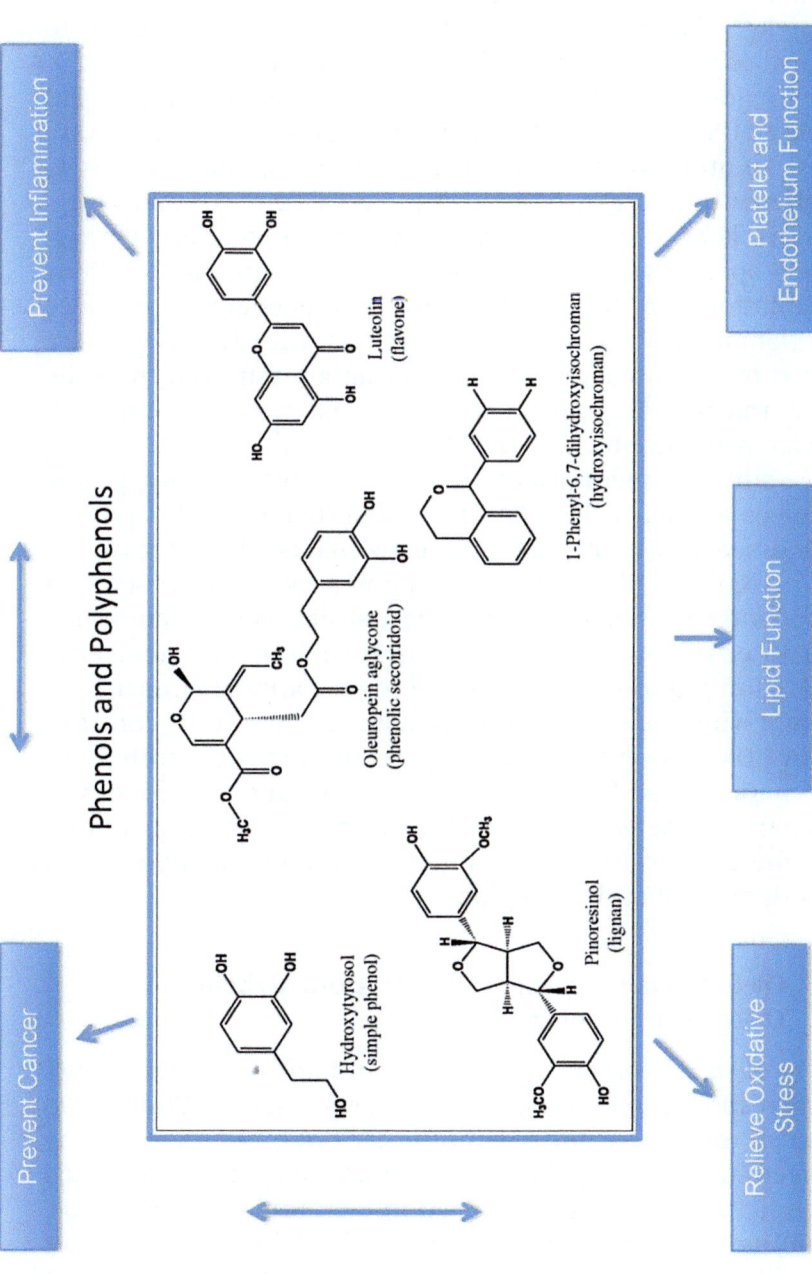

Figure 8.13 Multiple ways that phenols protect you (modified from Marie Josèph Amiot, Olive Oil And Health Effects: From Epidemiological Studies To The Molecular Mechanisms Of Phenolic Fraction, *OCL* **21**(5), D512, permission granted through Creative Commons Attribution License).

the triggering of genes that express enzymes to help deal with oxidative stress.[47] They are present in EVOO at trace levels (80–450 mg kg^{-1}) but a little goes quite a long way when it comes to chemical communication or signaling pathways.

The major antioxidants are based on hydroxytyrosol and include DHPEA-EDA and its cousin, which taste bitter. There are at least 50 other known antioxidants in EVOO, and it seems as though new ones are discovered every day. One of the many ways they work is as antioxidants that react with potentially damaging ROS species generated by other cellular processes such as inflammation and disease. The antioxidants chemically reduce those damaging ROS, and in the process, they themselves are changed into nonreactive molecules that are removed from the body. In that way, they act to sacrificially disarm the ROS that can cause so many problems in the body.

One additional factor that rounds out the health giving profile for molecules such as DHPEA-EDA is that their solubility profile, as discussed in Chapter 5, is perfectly balanced between solubility in aqueous phases and organic phases. In the arsenal of antioxidants accessible to our bodies, strong antioxidants such as the water-soluble vitamin C and the oil-soluble vitamin E exist but have limited solubility profiles. Even the antioxidants in fruit juice and wine are essentially only soluble in water. By contrast, DHPEA-EDA can get to almost all parts of the body – both those parts characterized by watery environments, such as blood, and those parts of the body, such as the brain, nervous tissue, or fat cells, where the environment is more hydrophobic. It acts as an antioxidant in all of these regions.

8.2.6 Oleocanthal: Fighting Inflammation, Alzheimer's Disease and Cancer

8.2.6.1 Fighting Inflammation. Inflammation is our body's natural response to trauma, infection, or cell death. White blood cells rush to the site and attempt to fix the problem by killing off infectious or damaged cells. Hormones called prostaglandins are made in response to an infection or trauma, and the prostaglandins cause increased fluids to collect in the infected or affected area that can cause swelling and pain. While this can be good for fighting a cold infection because it can help rid the body of infected or damaged material, it can also be bad if it is chronic and low level. Chronic inflammation can play a role in diseases such as arthritis, cancer, CVD, and even Alzheimer's disease.

For more than a century, aspirin was the most commonly used drug taken to relieve pain and inflammation, but its negative long-term effects on the gastrointestinal tract led researchers to look for other drugs. Steroids were discovered to be useful in combating swelling and inflammation, but they too have side effects that can be troubling. A new class of drugs, the non-steroidal anti-inflammatory drugs (NSAIDs), hit the market in the mid 80's as over the counter medications for the relief of pain. The active component of Advil, a popular NSAID, is ibuprofen, shown in Figure 8.14. Ibuprofen inhibits a pathway in the inflammatory response by binding to an enzyme and disrupting the signal for the production of the prostaglandins responsible for the swelling. For reasons that are most likely unrelated, but not known, ibuprofen produces a burning sensation in the back of your mouth – identical to the sensation produced by EVOO. Indeed, it was that observation that led Monel scientist Gary Beauchamp to link the back of the throat burning that he had felt in testing NSAIDs in his lab in Philadelphia with the same sensation he had when tasting olive oils while on vacation in Italy.[48] He subsequently confirmed what another scientist had reported earlier,[49] that olive oil contains oleocanthal (*oleo* from "olive" + *canth* for "stinging" + *al*dehyde for the particular chemically reactive group in the molecule), a substance that creates a burning sensation or pungency in the back of the throat. Not only does the oleocanthal, shown in Figure 8.14, create the same

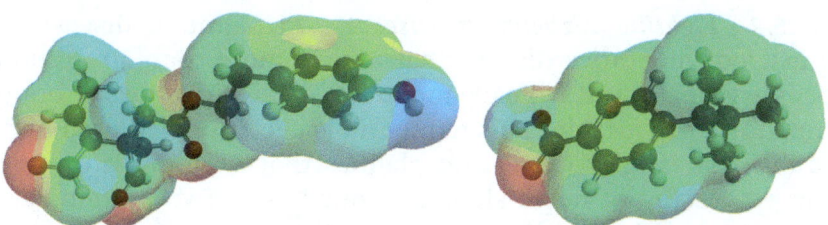

Figure 8.14 Two different ways to relieve pain using a non-steroidal anti-inflammatory drug (NSAID) – one from nature (oleocanthal, on left) and one from the pharmacist's laboratory (ibuprofen, on right).

sensation – it also acts to inhibit the inflammatory response in the same manner as the NSAID and so can be referred to as a natural NSAID.[50]

A robust "two or three cough" oil can contain 2.7 mg of oleocanthal per 50 g (about 3.5 T) of oil (0.0054% or 54 ppm).[51] While this is only about 10% of the dosage of one tablet of ibuprofen, daily consumption of this low level is likely to have protective properties similar to the daily consumption of baby aspirin recommended for many individuals at risk of CVD.[48] While such a finding is encouraging, biological effects or health giving properties of a natural substance such as olive oil rarely reduces to a single chemical.[52] With its association as an anti-inflammatory, it is not surprising to learn that individuals who incorporate EVOO into their diet are at lower risk of developing inflammatory diseases such as arthritis and hypertension and show some symptom relief by increasing the amount of EVOO in their diets. In Chapter 9, we discuss how, two thousand years ago, Olympic athletes would cover themselves with olive oil before and after an event – perhaps at one level, they were applying topical analgesic to help them deal with the aches and pains that would accompany extended strenuous physical activity.

The preventive and therapeutic role of oleocanthal in certain diseases is just beginning to be understood at a biochemical level. These studies have required basic understanding of the diseases at the level of molecular or cellular markers for the disease itself, and then the tools to be able to see how oleocanthal interrupts or changes the progression of the disease. Two such examples are discussed below.

8.2.6.2 Fighting Alzheimer's Disease. Alzheimer's disease is a neurodegenerative disease that afflicts more than 30 million people globally and typically manifests itself with a loss of brain function or dementia. This dementia is accompanied with neurofibrillary tangles (NFT) and plaques that slow down or stop the processing of nerve signals in our brain. Two proteins are associated with the disease, tau and β-amyloid protein (Aβ). The tau protein normally helps to stabilize neurons, but if the protein is slightly altered (either through genetic or chemical modification), the tau protein forms tangles. The Aβ is involved with many different cellular processes including the regulation of lipid metabolism. In patients with Alzheimer's disease, Aβ starts to start to stick together to form small assemblies which eventually become fibrous. They grow together to form plaque in diseased brain, slowing down function.[53]

When oleocanthal is added to tau protein, the formation of the fibrillary tangles is inhibited, as is shown in Figure 8.15 (left). When examined at a molecular level, it can be seen that the oleocanthal binds directly to tau protein, preventing its interaction with other tau proteins. The oleocanthal remains bound at the site, preventing the proteins from binding to other misfolded proteins and forming further plaques.[54] Another study also saw clearance of Aβ protein itself in cells in culture AND in mice brains when treated with oleocanthal, as shown in Figure 8.14 (right). In this case, the expression of genes for transport and other housekeeping proteins was enhanced.[55]

When we test potential drugs, we usually want to see that the amount of a drug is related to the strength of the effect. Notice that in Figure 8.15 the first bar shows how much disease is present in the untreated animal while subsequent bars show that the reduction of disease is stronger with more oleocanthal. We also want to have a plausible mechanism. Binding of the oleocanthal to the tau protein gives us that.

People whose diets are rich in EVOO show lowered risks of developing Alzheimer's disease. These studies suggest that the creation of these plaques and tangles is inhibited and/or breakdown and clearance of the plaques enhanced once they form. Research is ongoing for ways to utilize these observations in designing therapeutic targets against diseases caused by abnormal protein folding and aggregation (amyloidopathies).[56]

8.2.6.3 Fighting Breast Cancer. Breast cancer is the most common form of cancer among women worldwide, with 1.7 million new cases being reported in 2012.[57] A woman stands a one in eight chance of developing breast cancer in her lifetime. The epidemiological studies show a Mediterranean Diet reduces your risk of invasive breast cancer by as much as 20% in postmenopausal women.[58] How does this work?

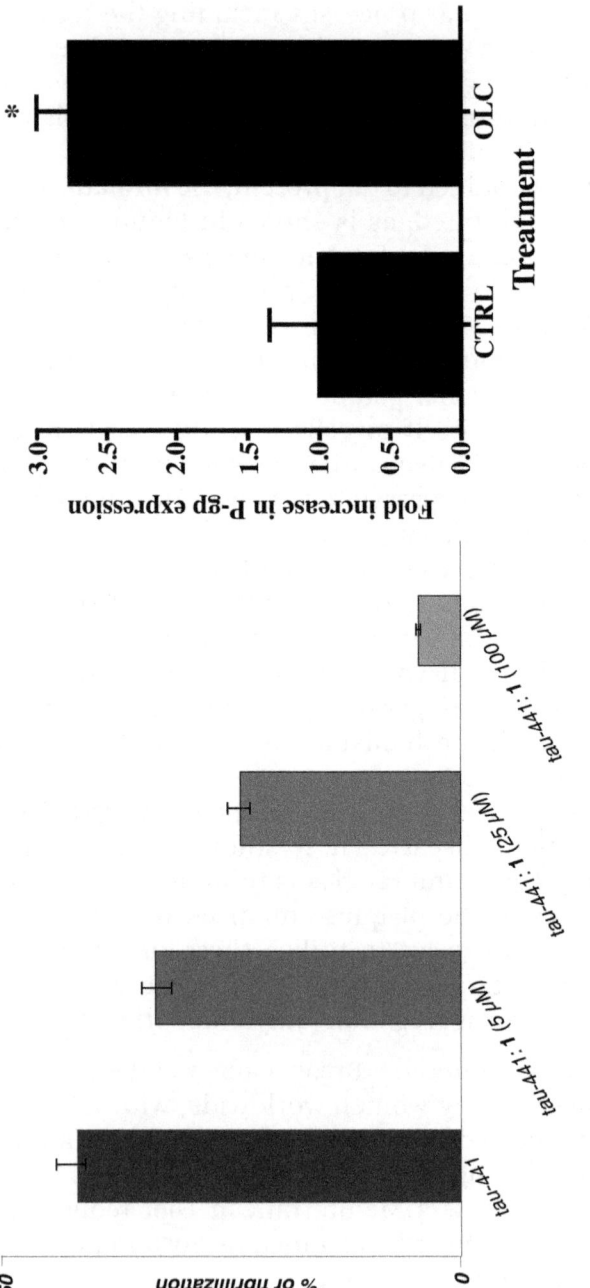

Figure 8.15 Oleocanthal protects against Alzheimer's disease in at least two ways. The left panel shows how increasing levels of oleocanthal cause reduced levels of tau protein that lead to formation of fibrils in the brain.[54] Adapted with permission from Monti, *J. Nat. Prod.* **75**(9), 1584–1588. Copyright 2012 American Chemical Society. The right panel shows that, compared with control mouse brains (CTRL), brains treated with oleocanthal (OLC) show 3.5× greater production of a transporter protein, P-ab, that helps to clear excess Aβ protein and prevent fibril formation.[55] Adapted with permission *ACS Chemical Neuroscience*, American Chemical Society, June 1 2013. Copyright 2013 ACS.

At least two proteins important for breast cancer cell growth are affected by oleocanthal. One of them, c-Met is a signaling protein in the membrane – and when there is a lot of this protein, cancer cells will be induced to grow in a malignant fashion (uncontrolled and unregulated). Oleocanthal has been shown to inhibit this protein in a dose dependent way. Computer simulations suggest that the oleocanthal actually binds to the protein at the active binding site, as shown in Figure 8.16. As a result, scientists are using oleocanthal as a therapeutic target starting point for developing drugs that will help to prevent the signaling protein from causing the cancer to spread.[59]

One thing to know about cancer cells is that they grow at a rate that far exceeds that of normal cells. Because of this, they use lots of fuel and lots of oxygen and produce lots of cellular waste (like garbage) that is stored in vessels called lysosomes inside the cancer cells. These vessels contain many enzymes that are toxic to cells as well as nasty byproducts, enzymes that might be damaged or oxidized, and other waste products of the cell. They are kept in this vacuole like your trash is kept in a trash bag, until the time at which the cell might be able to empty the trash or until the time the cell dies. When oleocanthal is

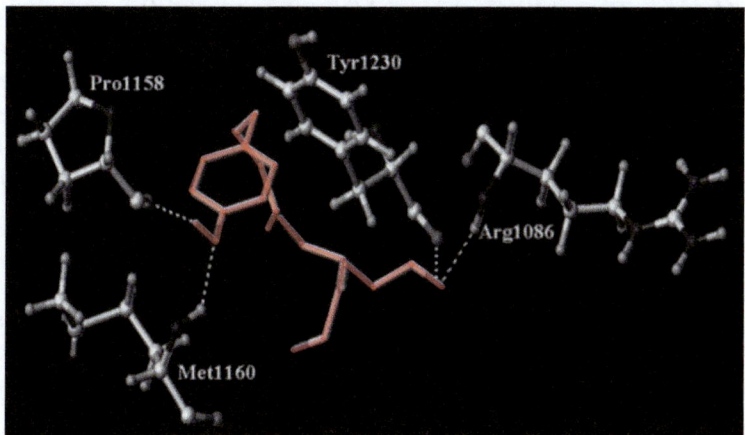

Figure 8.16 Computational model of oleocanthal binding to the met-c protein and inhibiting its activity.[59] Reprinted with Permission from © Georg Thieme Verlag KG.

added to breast cancer cells that have been cultured in a laboratory, the breast cancer cells die very quickly – within 30 minutes or so. Before the cells die, the lysosomes within the cells seem to have dissolved, spilling the toxic contents out into the cell – which would quickly trigger cell death as sure as would a leak of a toxic chemical into the environment. The rapid death of the breast cancer cells could be explained by oleocanthal's entering the membrane of the vacuole and poking a hole in it, causing the bag to break and the contents to spill out. Normal cells did not show this rapid response to the oleocanthal. New research has shown that an enzyme that regulates the permeability of these lysosomes in cancer cells, acid sphingomyelinase, has its activity altered by oleocanthal.[60] Perhaps exposing cancer cells early to oleocanthal (and to the antioxidants in the EVOO) prevents the spread of the disease before it becomes a problem.

While the answers to how EVOO acts to fight cancer are not straightforward or simple, neither is this disease. Researchers at the Universitat Autònoma de Barcelona, Spain, in the Department of Cell Biology, Physiology and Immunology and the Medical Physiology Unit, Medicine School, summarized the current state of affairs.[61]

"The specific mechanisms by which EVOO and other dietary lipids may exert their modulatory effects on cancer are not fully understood although abundant research has proposed the following: They influence in the stages of the carcinogenesis process, oxidative stress, alteration of the hormonal status, modification of the structure and function of cell membranes, modulation of cell signaling transduction pathways, regulation of gene expression and influence in the immune system."

Again, population studies have shown that consumption of EVOO is associated with a lower risk of cancer – in particular breast cancer and colon cancer. While today, our methods of running these studies are quite sophisticated, statistically it should be noted that the preventive role of EVOO in the fight against breast cancer was first proposed in 1614 by Giacomo Castelvetro who believed in the "Sacred Law of Salad" and recommended a diet of greens, raw vegetables, and olive oil.[62]

8.3   A WORLD OF OTHER HEALTH BENEFITS

Consumption of EVOO has been linked to the prevention and treatment of many other diseases including osteoporosis, arthritis, skin diseases, wound healing, depression, and migraine headaches. It is likely that the exact details of these interactions will be a wonderful interplay of the themes we have developed here: providing our body with a rich energy dense nutrient that helps balance the nutritive needs of our body while helping to prevent obesity by keeping us full and satisfied, supplying powerful antioxidants that have solubility profiles that make them reach places in our body not accessible to other antioxidants, preventing inflammation through the ability of oleocanthal to inhibit inflammatory pathways, and in the complex regulation of genes, hormones, our microbiome, and the general homeostasis that helps to insure a healthy life.

In the 10th Century, Al-Razi advised the Islamic world "If you can cure a person by diet, do not suggest medication.[63]" After all is said and done, we hope you will be convinced that olive oil used in moderation (3–4 tablespoons per day) and in combination with other Mediterranean foods will go a long way towards helping you realize good health and a long and happy life – perhaps not the 122 years Jeanne Louise Calment enjoyed, but who are we to say?

REFERENCES

1. *World's Oldest Person Dies*, https://news.google.com/news-papers?id=VqEgAAAAIBAJ&sjid=7GgFAAAAIBAJ&pg=602633 13956&hl=en, accessed June 2016.
2. D. de Materia Medica, *Being an Herbal with many other Medicinal Materials Written in Greek in the First Century of the Commob Era a new Indexed Version in Modern English by ta Osbaldeston and Rpa Wood*, 2000.
3. *The Canon of Medicine*, http://ddc.aub.edu.lb/projects/saab/avicenna/contents-eng.html.
4. M. Nasser, A. Tibi and E. Savage-Smith, *J. R. Soc. Med.*, 2009, **102**, 78.
5. I. Q. al-Jawziyya, *Penelope Johnstone (Cambridge: Islamic Texts Society, 1998)*, 1998, p. 80.

6. *What is Wrong with Olive Oil*, https://www.pritikin.com/your-health/healthy-living/eating-right/1103-whats-wrong-with-olive-oil.html, accessed June 2016.
7. V. Tremaroli and F. Bäckhed, *Nature*, 2012, **489**, 242.
8. H. J. Flint, *J. Clin. Gastroenterol.*, 2011, **45**, S128.
9. J. R. Mujico, G. C. Baccan, A. Gheorghe, L. E. Díaz and A. Marcos, *Br. J. Nutr.*, 2013, **110**, 711.
10. J. F. Brinkmann, N. A. Abumrad, A. Ibrahimi, G. J. van der Vusse and J. F. Glatz, *Biochem. J.*, 2002, **367**, 561.
11. J. M. Berg, J. L. Tymoczko and L. Stryer, *Biochemistry*, W. H. Freeman, New York, 2002.
12. C. K. Chow, *Fatty Acids in Foods and Their Health Implications*, CRC Press, Boca Raton, 2008.
13. *Essential Fatty Acids; Intake Requirement*, http://www.nutri-facts.org/eng/essential-fatty-acids/essential-fatty-acids/intake-recommendations/, accessed June 2016.
14. M. D. Klok, S. Jakobsdottir and M. L. Drent, *Obes. Rev.*, 2007, **8**, 21.
15. N. Sáinz, C. J. González-Navarro, J. A. Martínez and M. J. Moreno-Aliaga, *Expert Opin. Ther. Targets*, 2015, 1.
16. E. Rozengurt and C. Sternini, *Curr. Opin. Pharmacol.*, 2007, **7**, 557.
17. S. Janssen, J. Laermans, P. J. Verhulst, T. Thijs, J. Tack and I. Depoortere, *Proc. Natl. Acad. Sci. U. S. A.*, 2011, **108**, 2094.
18. R. J. Seeley and S. C. Woods, *Nat. Rev. Neurosci.*, 2003, **4**, 901.
19. M. J. Amiot, *OCL*, 2014, **21**, D512.
20. E. Ros, M. A. Martínez-González, R. Estruch, J. Salas-Salvadó, M. Fitó, J. A. Martínez and D. Corella, *Adv. Nutr.*, 2014, **5**, 330S.
21. A. Alonso, V. Ruiz-Gutierrez and M. A. Martínez-González, *Public Health Nutr.*, 2006, **9**, 251.
22. I. Zazpe, A. Sánchez-Tainta, E. Toledo, A. Sanchez-Villegas and M. Martínez-González, *J. Acad. Nutr. Diet.*, 2014, **114**, 37.
23. C. Samieri, C. Féart, C. Proust-Lima, E. Peuchant, C. Tzourio, C. Stapf, C. Berr and P. Barberger-Gateau, *Neurology*, 2011, **77**, 418.
24. B. Bendinelli, G. Masala, C. Saieva, S. Salvini, C. Calonico, C. Sacerdote, C. Agnoli, S. Grioni, G. Frasca and A. Mattiello, *Am. J. Clin. Nutr.*, 2011, 275.

25. G. Buckland, A. L. Mayén, A. Agudo, N. Travier, C. Navarro, J. M. Huerta, M. D. Chirlaque, A. Barricarte, E. Ardanaz, C. Moreno-Iribas, P. Marin, J. R. Quirós, M. L. Redondo, P. Amiano, M. Dorronsoro, L. Arriola, E. Molina, M. J. Sanchez and C. A. Gonzalez, *Am. J. Clin. Nutr.*, 2012, **96**, 142.
26. J. E. Dalen and S. Devries, *Am. J. Med.*, 2014, **127**, 364.
27. S. Jeremiah, J. D. Neaton, J. D. Cohen, J. Cutler, L. Eberly, G. Grandits, L. H. Kuller, J. Ockene, R. Prineas and The MRFIT Research Group, *J. Am. Heart Assoc.*, 2012, **1**, e003640.
28. J. Stamler, J. D. Neaton, J. D. Cohen, J. Cutler, L. Eberly, G. Grandits, L. H. Kuller, J. Ockene, R. Prineas and MRFIT Research Group, *J. Am. Heart Assoc.*, 2012, **1**, e003640.
29. A. Hruby and F. B. Hu, *Lipid Technol.*, 2016, **28**, 7.
30. R. Chowdhury, S. Warnakula, S. Kunutsor, F. Crowe, H. A. Ward, L. Johnson, O. H. Franco, A. S. Butterworth, N. G. Forouhi, S. G. Thompson, K. T. Khaw, D. Mozaffarian, J. Danesh and E. Di Angelantonio, *Ann. Intern. Med.*, 2014, **160**, 398.
31. R. J. de Souza, A. Mente, A. Maroleanu, A. I. Cozma, V. Ha, T. Kishibe, E. Uleryk, P. Budylowski, H. Schünemann, J. Beyene, *Br. Med. J.*, 2015, 351.
32. M. de Lorgeril, P. Salen, J. L. Martin, I. Monjaud, J. Delaye and N. Mamelle, *Circulation*, 1999, **99**, 779.
33. P. Kris-Etherton, R. H. Eckel, B. V. Howard, S. St Jeor, T. L. Bazzarre and Nutrition Committee Population Science Committee and Clinical Science Committee of the American Heart Association, *Circulation*, 2001, **103**, 1823.
34. M. Guasch-Ferré, F. B. Hu, M. A. Martínez-González, M. Fitó, M. Bulló, R. Estruch, E. Ros, D. Corella, J. Recondo, E. Gómez-Gracia, M. Fiol, J. Lapetra, L. Serra-Majem, M. A. Muñoz, X. Pintó, R. M. Lamuela-Raventós, J. Basora, P. Buil-Cosiales, J. V. Sorlí, V. Ruiz-Gutiérrez, J. A. Martínez and J. Salas-Salvadó, *BMC Med.*, 2014, **12**, 78.
35. U. S. D. O. Health and H. Services, *2015–2020 Dietary Guidelines for Americans*, 2015.
36. G. D. Lopaschuk, J. R. Ussher, C. D. Folmes, J. S. Jaswal and W. C. Stanley, *Physiol. Rev.*, 2010, **90**, 207.
37. R. Lahey, X. Wang, A. N. Carley and E. D. Lewandowski, *Circulation*, 2014, **130**, 1790.
38. A. A. R. Al-Shudiefat, A. K. Sharma, A. K. Bagchi, S. Dhingra and P. K. Singal, *Mol. Cell. Biochem.*, 2013, **372**, 75.

39. R. J. Widmer, A. J. Flammer, L. O. Lerman and A. Lerman, *Am. J. Med.*, 2015, **128**, 229.
40. *Cholesterol Ratio: Is it Important?*, http://www.mayoclinic.org/cholesterol-ratio/expert-answers/faq-20058006, accessed June 2016.
41. V. Llorente-Cortés, R. Estruch, M. P. Mena, E. Ros, M. A. M. González, M. Fitó, R. M. Lamuela-Raventós and L. Badimon, *Atherosclerosis*, 2010, **208**, 442.
42. R. Carnevale, P. Pignatelli, C. Nocella, L. Loffredo, D. Pastori, T. Vicario, A. Petruccioli, S. Bartimoccia and F. Violi, *Atherosclerosis*, 2014, **235**, 649.
43. R. Carnevale, L. Loffredo, M. Del Ben, F. Angelico, C. Nocella, A. Petruccioli, S. Bartimoccia, R. Monticolo, E. Cava and F. Violi, *Clin. Nutr.*, 2016, DOI: 10.1016/j.clnu.2016.05.016.
44. R. Estruch, M. A. Martínez-González, D. Corella, J. Salas-Salvadó, M. Fitó, G. Chiva-Blanch, M. Fiol, E. Gómez-Gracia, F. Arós, J. Lapetra, L. Serra-Majem, X. Pintó, P. Buil-Cosiales, J. V. Sorlí, M. A. Muñoz, J. Basora-Gallisá, R. M. Lamuela-Raventós, M. Serra-Mir, E. Ros and PREDIMED Study Investigators, *Lancet Diabetes Endocrinol.*, 2016, 666.
45. H. E. Bloomfield, E. Koeller, N. Greer, R. MacDonald, R. Kane and T. J. Wilt, *Ann. Intern. Med.*, 2016, **65**, 491.
46. A. Bergouignan, I. Momken, D. A. Schoeller, C. Simon and S. Blanc, *Prog. Lipid Res.*, 2009, **48**, 128.
47. H. J. Forman, K. J. Davies and F. Ursini, *Free Radicals Biol. Med.*, 2014, **66**, 24.
48. G. K. Beauchamp, R. S. Keast, D. Morel, J. Lin, J. Pika, Q. Han, C. H. Lee, A. B. Smith and P. A. Breslin, *Nature*, 2005, **437**, 45.
49. P. A. Breslin, T. N. Gingrich and B. G. Green, *Chem. Senses*, 2001, **26**, 55.
50. S. Cicerale, P. A. Breslin, G. K. Beauchamp and R. S. Keast, *Chem. Senses*, 2009, **34**, 333.
51. L. Parkinson and R. Keast, *Int. J. Mol. Sci.*, 2014, **15**, 12323.
52. V. Fogliano and R. Sacchi, *Mol. Nutr. Food Res.*, 2006, **50**, 5.
53. H. Lei, *Protein Cell*, 2010, **1**, 312.
54. M. C. Monti, L. Margarucci, R. Riccio and A. Casapullo, *J. Nat. Prod.*, 2012, **75**, 1584.
55. A. H. Abuznait, H. Qosa, B. A. Busnena, K. A. El Sayed and A. Kaddoumi, *ACS Chem. Neurosci.*, 2013, **4**, 973.
56. A. Attar and G. Bitan, *Curr. Pharm. Des.*, 2014, **20**, 2469.

57. *Breast Cancer Statistics*, http://www.wcrf.org/int/cancer-facts-figures/data-specific-cancers/breast-cancer-statistics, accessed June 2016.
58. G. Buckland, N. Travier, V. Cottet, C. A. González, L. Luján-Barroso, A. Agudo, A. Trichopoulou, P. Lagiou, D. Trichopoulos, P. H. Peeters, A. May, H. B. Bueno-de-Mesquita, F. J. Bvan Duijnhoven, T. J. Key, N. Allen, K. T. Khaw, N. Wareham, I. Romieu, V. McCormack, M. Boutron-Ruault, F. Clavel-Chapelon, S. Panico, C. Agnoli, D. Palli, R. Tumino, P. Vineis, P. Amiano, A. Barricarte, L. Rodríguez, M. J. Sanchez, M. D. Chirlaque, R. Kaaks, B. Teucher, H. Boeing, M. M. Bergmann, K. Overvad, C. C. Dahm, A. Tjønneland, A. Olsen, J. Manjer, E. Wirfält, G. Hallmans, I. Johansson, E. Lund, A. Hjartåker, G. Skeie, A. C. Vergnaud, T. Norat, D. Romaguera and E. Riboli, *Int. J. Cancer*, 2013, **132**, 2918.
59. A. Y. Elnagar, P. W. Sylvester and K. A. El Sayed, *Planta Med.*, 2011, **77**, 1013.
60. O. LeGendre, P. A. S. Breslin and D. A. Foster, *Mol. Cell. Oncol.*, 2015, e1006077.
61. E. Escrich, M. Solanas, R. Moral and R. Escrich, *Curr. Pharm. Des.*, 2011, **17**, 813.
62. R. Colomer, R. Lupu, A. Papadimitropoulou, L. Vellón, A. Vázquez-Martín, J. Brunet, A. Fernández-Gutiérrez, A. Segura-Carretero and J. A. Menéndez, *Clin. Transl. Oncol.*, 2008, **10**, 30.
63. *Food as Medicine in Muslim Civilization*, http://www.muslim-heritage.com/article/food-medicine-muslim-civilization, accessed June 2016.

CHAPTER 9

1001 Uses for Olive Oil

There are few materials with which humans have had such a long, unbroken, and almost unchanged relationship as they have with olive oil. The fossil record shows that olive trees and humans have coexisted for at least 45 000 years.[1] The uses of olive oil are as wide and varied as that long history might suggest. During the contact with the olive tree and its products, the application of human creativity has resulted in uses from birth to death, from athletic activity to holy worship, from basic nutrition to nutra-ceuticals – it is hard to find any aspect of human civilization in which olive oil has *not* played a role at some time or another.[2]

We imagine the people who first discovered that olives were edi-ble. What a leap of imagination or observation it took to discover that this extremely bitter, unpleasant little fruit can become deli-cious if soaked in salt water for a few weeks! What human tens of thousands of years ago could notice the separation of oil and water in the juice from ripe olives, and isolate the oil to save, discard-ing the dark bitter vegetable water just as we do today? That same inventiveness and talent for close observation created uncounted uses for the oil, and for other products of the olive tree.

We also imagine the vast benefits that olive oil has brought to human society over the millennia of contact. Far from limited to use as a food, it is useful in contact with nearly all parts of the

The Chemical Story of Olive Oil: From Grove to Table
By Richard Blatchly, Zeynep Delen Nircan and Patricia O'Hara
© Richard Blatchly, Zeynep Delen Nircan and Patricia O'Hara, 2017
Published by the Royal Society of Chemistry, www.rsc.org

body at all ages (and even after death according to some traditions). As a general lubricant, it can ease the turning of ox-cart wheels or other machines. In lamps, it burns with a beautiful white light. Even the soot created above the lamp is useful. It is easy to see why ancient Greeks thought that this tree was one of the greatest gifts from the gods.

9.1   IN THE KITCHEN

One of the simplest ways that extra virgin olive oil (EVOO) is useful is that it makes healthy foods, such as salads and vegetables, taste better – and better tasting food makes you eat more. Any greens – even the most bitter leaf vegetable or tasteless legume will be livened up with a dash of your favorite EVOO.

Matching the right EVOO with the food you are eating can be a bit difficult in the beginning, until you identify the right oil for the right food. To help you with these decisions, the New York International Olive Oil Competition (NYIOOC) has developed an oil–food pairing app. Here is what they say about this:

> " *With more than one thousand olive cultivars and countless taste profiles among them, finding the best match for a certain food can be a daunting challenge. Not anymore ... a food pairing app [has been developed] to identify the best match for your culinary creation among this year's award-winning oils. "We know that a great olive oil can make foods soar," said Curtis Cord, the NYIOOC president, and Olive Oil Times publisher, "but choosing the wrong one can diminish the result. This new tool takes the guesswork out of finding the best match."*[3]

While EVOO can be consumed drizzled on vegetables or shaken on salads, it is often used as a way of sautéing or frying meats or vegetables. Many foods need to be cooked in order to be digestible. Water is often the major ingredient, especially of plants, and when heated it turns into steam and breaks down cell walls. A chewy, tough vegetable – potato, turnip, parsnip – can be transformed into something soft and palatable. Protein containing foods, such as eggs or chicken, will also be transformed when heated, in a process known as denaturation. Since all protein must be denatured in order to be digested, cooking accelerates the digestive process. One additional benefit of cooking our foods was especially important to our ancestors, who needed to make foods last from one harvest or hunt to another and yet lacked refrigerators or the tools to preserve foods. Without modern hygiene, stored foods would collect food borne pathogens that are killed by heating. Even today, in general, freshly cooked foods are safer than raw food, and cooking with EVOO is a great way of making your food taste even better!

When using EVOO to sauté vegetables instead of cooking them with water, you are also providing an added benefit – locking in flavors and health giving substances from the vegetables that might otherwise leach out into the water if your substance is boiled, poached, or steamed. For example, the Brassica family of vegetables, which includes broccoli, cauliflower, Brussels sprouts, turnips, mustards, cabbage, and kale, is known to be comprised of healthy foods that protect you against certain diseases, such as colon and rectal cancer.[4] The active ingredients in these vegetables are glucosinolates, shown in Figure 9.1, a powerful class of antioxidants that, unlike the antioxidants in EVOO, are water-soluble. Cooking with EVOO locks all of those great cancer-protecting groups inside your vegetables rather than allowing them to be diluted into the heating water. A cup of Brussels sprouts prepared

Figure 9.1   The antioxidants known as glucosinolates come from Brussels sprouts.

by sautéing or roasting in EVOO gives you a double dose of antioxidants – both the water-soluble vegetable antioxidants *and* the oil-soluble antioxidants from the EVOO.

In addition, recent studies have shown that some of the anti-oxidants found in olive oil are actually transferred to a variety of vegetables if they are sautéed in the oil. Similar to the behavior of Brassica vegetables, these vegetables also retain their natural antioxidants better. Neither of these benefits are found in vege-tables that have been boiled or cooked in an oil–water mixture.[5]

Tomatoes are famously healthy for us because they contain lyco-pene, shown in Figure 9.2, a molecule in the carotenoid family that is responsible for the bright red color of the vegetable. The health benefits of lycopene are similar to those of EVOO: pre-vention of cardiovascular disease, aiding digestion, and helping to prevent cancer through its antioxidant properties. Lycopene in tomatoes exists in two different forms: *cis*-lycopene and *trans*-lycopene. Chemists call these different forms isomers. Only the *cis* isomer is readily absorbed by our bodies, yet the prevalent form in uncooked tomatoes is the *trans* isomer. Cooking helps convert the *trans* to the *cis*, thus making the health benefits more available to our bodies. Studies have shown that cooking tomatoes in EVOO is an efficient way of improving the bioavailability of lycopene.[6–8]

Cooking with rosemary and olive oil together can have syner-gistic effects. Rosemary contains important antioxidants carno-sol, carnosic acid, and rosmarinic acid. These compounds have diverse biological activity including being anti-inflammatory, anti-viral, cytoprotective, and anti-tumoral. The active com-pounds work as antimicrobial agents to work against *Listeria* strains.[9] Extracts of rosemary have even been shown to improve

Figure 9.2 The antioxidant lycopene found in raw tomatoes. The lycopene is converted from one isomer (*trans*) into another isomer (*cis*) when cooked in olive oil. The *cis* isomer is more readily absorbed by our bodies than the *trans*.

spatial memory in rats.[10] Since the bioactive molecules are oil-soluble, infusing oil with rosemary, or simply eating them together, can make these healthy molecules more bioavailable.

Sometimes, cooking should just be for fun, and olive oil taken to very low temperatures will afford quite a whimsical confection – an olive oil lollipop. With a plate or metal surface that has been super-chilled, it is possible to easily whip up a batch of these amazingly simple confections. We chilled our metal cooking surface (the bottom of a metal pot) with liquid nitrogen, but if you don't have access to that, dry ice will do. Slowly and carefully pour out 1 T of the oil onto the chilled surface (not too thin!) and insert a lollipop stick when the oil close to the metal surface becomes opaque. Use a metal spatula to flip the lollipop to finish off the hardening on the opposite surface. Remove from the surface and enjoy. The buttery mouth feel of the chilled oil is lovely, and when the oil melts and reveals its full essence, the blossoming flavor is divine. We accented the pops with both a sweet/hot (chocolate, sugar, chili pepper) and savory (herb mixture) additives. Trust your own imagination for new pairings.

Cameo 9.1   Argyros Tzanakis, Kritsa, Greece.

We met Argyros Tzanakis one night in mid December in the tiny Cretan town of Kritsa. Navigating the twisty roads in the dark searching for our hotel had been tricky and we were tired. What a relief to see the well-lit sign for Argyros Guest House! Pulling into the narrow driveway we were startled by a jubilant "Bravo! Bravo!" from someone who clearly had been awaiting our arrival. We soon realized this was Argyros herself waving for us to pull in close to the main house. Grabbing our bags and marching us up to our rooms, she made us feel welcome and well cared for.

Later, over tea and cakes, she told us that Kritsa is surrounded by mountains and, in these mountains, one can find the Katharo plateau up at 1200 m with views to the sea. In early summer, sheep are herded onto flower filled fields and local shepherds and villagers tend them throughout the summer. Argyros remembers summers spent in the plateau with her grandmother, learning how to care for the animals and farm and harvest various vegetables, collect honey, make yogurt and cheese. In between, she taught herself English from books and television. Today, Argyros runs this small guest house, cooks breakfast and dinner for guests at the inn, farms a small plot of land close to the village, takes care of the chickens, goats, and several dogs, and now, in the harvest season, is out harvesting olives from the family grove of 500 trees. She works with her grown son who is taking time off from his job as a police officer in a nearby town. The olives are harvested with a motorized hand held rake that combs through the trees to shake off the ripe Koroneiki olives into a net spread on the ground. The two of them will harvest perhaps 50 trees in a day. By evening, the olives are collected into burlap bags, taken to the local cooperative press, weighed, washed, and on their way to becoming prize winning Kritsa Olive Oil. As the purpose of our visit is to learn about the olive harvest, we are allowed to "help" with the harvest. After a 30 minute lesson, we stop for a field snack of fresh walnuts, cheese, and a homemade liquor called Grappa.

Argyros will save several kilos of olives to make her family's delicious table olives. She served us some for breakfast with honey harvested from hives in her garden on yogurt she has cultured. Her house is spread with brilliant rugs woven by her grandmother, stone mills and mortars used to do the pressing a millennium ago, and a basket of bright orange clementines that ripen alongside the olives in groves on the hillside.

Argyros website: http://www.argyrorentrooms.gr/index.html.

9.1.1 Cooking with Flavored Oils

While, technically speaking, the addition of anything to an oil means that the oil can no longer be classified as extra virgin, the growing popularity of flavored oils merits mention here. Some really delicious oils can be prepared in this way. If you've never had the opportunity to dress a salad with a lemon-flavored oil, you are missing something.

Flavored oils can be prepared by adding flavoring extracts to an extra virgin or virgin olive oil or through a process known as co-milling, where the fruit, herb, or spice is added in with the olives during the crush. If an extract is used, the quality of the product will likely depend upon the quality of the extract. Like vanilla extract you use in baking, these extracts can be either artificial or natural and one can pay a pretty penny for some of the best extracts.

We heard of a producer who was experimenting with co-milling Meyer lemons with olives using a stone mill. Thinking that the taste would be improved with the inclusion of the whole fruit, he dumped several boxes of ripe lemons into the hopper with his olives. The crushing process was going along nicely and a layer of oil was building up on top of the olives. The miller turned away to adjust the controls and felt an object whizz past his head at great speed and make a "splat" against the far wall. As he turned back toward the mill, he saw that conditions were now right for the still intact, but now greased up, lemons to be shot out from the millstone like ammunition from a cannon. He ducked his way to the controls to turn off the mill but not before several dozen more lemon cannonballs were shot out of the stone mill. He now cuts his lemons in half with excellent results.

One of our favorite recipes using flavored oils is for Orange Brownies with sea salt, and was sent to us by Marvin Martin, founder of Martin Olive Oils near Sacramento and chef at the Mondavi Winery. The recipe uses no butter and only a 1/4 cup of a mandarin orange flavored olive oil. The marriage of the orange flavor with the chocolate and the sea salt is heavenly. You can find this recipe and more on our blog (http://worldolivepress. blogspot.com).

9.1.2 Safe Cooking with EVOO – Keep it Below the Smoke Point

A concern often heard is that EVOO is not "safe" for cooking. Some say that heating EVOO even at a fairly low temperature breaks down the oil and other components and creates dangerous byproducts. Perhaps thinking about this scientifically will put these concerns in their place.

The temperature at which a heated oil begins to smoke continuously is called the smoke point. No oil should be used for cooking at or above its smoke point for two reasons. First, the breakdown products have off flavors and can be unhealthy and second, the next step (though not a small one) from a smoking oil is a flaming oil, and no one wants to burn down their kitchen. At the smoke point, the triacylglycerides (TAGs) break down first to their component fatty acids and glycerol, and then these compounds themselves are further degraded. Some of the products vaporize to create a bluish smoke, as shown in Figure 9.3. One component in the smoke is acrolein, a molecule derived from glycerol that is acrid smelling and irritating to your eyes and throat. Acrolein, as shown in the left panel of Figure 9.3, belongs to a reactive class of molecules which have been shown to be carcinogenic. Since the breakdown products (including free fatty acids, FFAs) accumulate in the heated oil over time, and since the increasing FFA acid lowers the smoke point, one should really avoid as much as possible reusing oil for frying.

The smoke point is unique to each oil and depends on many factors, including the identity of the fatty acids that make up the TAGs and the amount of FFAs in the oil. Several examples are given in Table 9.1. Plant oils with saturated fats (coconut and palm kernel oils) have very low smoke points because they contain low molecular weight fats and, for that reason, they should not be used for frying. Butter contains saturated fats and some milk proteins, and can decompose rather quickly when heat is added. Most plant and seed oils are liquid at room temperature, and when not chemically processed, may contain low levels of FFAs. The FFA molecules act as catalysts to increase the rate of

Figure 9.3   The acrolein breakdown product of olive oil produced at the smoke point is shown at the left and at the right is a picture of oil at the smoke point.

Table 9.1 Smoke points of several cooking oils.[a]

Source	ω6:ω3 (% FA)	Smoke point °F/°C		
		Unrefined (natural)	Refined (chemical)	Other
Flaxseed (linseed)	1:4	225 °F/107 °C		
%SFA	11.0			
%MUFA	18.5			
%PUFA	70.5			
Safflower	133:1	225 °F/107 °C	510 °F/266 °C	320 °F/160 °C semirefined
%SFA	6.5			
%MUFA	15.1			
%PUFA	78.4			
Corn	83:1	320–352 °F/ 160–178 °C	450 °F/232 °C	
%SFA	13.6			
%MUFA	29.0			
%PUFA	57.4			
Butter	9:1	250–350 °F/ 121–177 °C		
%SFA	68.0			
%MUFA	27.9			
%PUFA	4.1			
Peanut	32:1	320 °F/160 °C	450 °F/232 °C	
%SFA	17.8			
%MUFA	48.6			
%PUFA	33.6			
Vegetable (soy)	7.5:1		466 °F/241 °C	
%SFA	16.3			
%MUFA	23.7			
%PUFA	60.0			
Lard		370 °F/182 °C		
%SFA	41.1			
%MUFA	47.2			
%PUFA	11.7			
Canola[b]	3:1; 2:1		468 °F/242 °C	400 °F/200 °C expeller processed
%SFA	7.4			
%MUFA	64.1			
%PUFA	28.5			
Olive	*13:1*	*405 °F/207 °C high quality extra virgin 0.2% FFA*	*460 °F/238 °C refined*	*468 °F/242 °C light (mixed refined and virgin)*
%SFA	*14.2*	*374 °F/190 °C extra virgin 0.5% FFA*		
%MUFA	*75.0*	*347 °F/175 °C virgin 0.8% FFA*		
%PUFA	*10.8*	*320 °F/160 °C virgin 2% FFA*		

Avocado oil[c] 12:1		520 °F/271 °C
%SFA	12.1	
%MUFA	73.8	
%PUFA	14.1	

[a]Adapted from http://jonbarron.org/diet-and-nutrition/healthiest-cooking-oil-chart-smoke-points#.VXA8SkvC_tt and Wikipedia http://en.wikipedia.org/wiki/Smoke_point.
[b]Hybrid plant bred from rapeseed developed in 1970 in Canada.
[c]Avocado oil has such a high smoke point because it is so low in FFA, also high in vitamin E. As per Marie Wong from Massey University, New Zealand (email communication).

breakdown of the oil and, as the temperature rises, the rate of the breakdown reaction increases until the point at which there is sufficient breakdown to produce smoke. The FFA level is probably the single most important factor when comparing liquid oils.

Now, what about olive oil? Olive oil comes in many grades, and each of the grades has a different upper limit for FFAs. IOC standards require oils classified as extra virgin to have <0.8% FFA, virgin olive oil <2% FFA, and "ordinary" – or sometimes confusingly named "pure" – olive oil <4% FFA. Chemically refined oil has no FFAs because solvents have been used to "purify" the oil, removing antioxidants, flavors, and any FFAs. Within each grade exists quite a spread of allowed levels of FFAs, which is why two "extra virgin" olive oils may have very different smoke points. A higher quality EVOO will have a lower FFA and thus a higher smoke point. Re-using oil for frying will also increase the FFA and lower the smoke point. When one uses high quality EVOO there should be no problem with it breaking down under normal cooking conditions. Alternatively, a refined oil can be used for cooking, and finished with a drizzle of EVOO, but you will then lose out on the added antioxidants that the EVOO provides.

Dr Mary Flynn, Adjunct Professor at Brown University and physician, speaks from both her own research and that of others[11–13] when she makes the following comments about cooking with EVOO:

"Nearly every time I lecture on olive oil, people ask whether heat destroys the oil, and whether they can cook with it. I don't know who invented this misconception (seed-oil companies?), but I'd love to dispel it once and for all. High quality extra virgin olive oil can be heated to 420 °F (215 °C) before it reaches smoke-point (i.e. oil smokes and starts to form unhealthy compounds), which is

higher than nearly every other vegetable oil. Olive oil is much more stable when heated compared to most vegetable oil. Cooking with olive oil below the smoke-point does not destroy most of its health benefits, or make it less healthy – under normal cooking conditions, most of the therapeutic minor components are retained".[14]

Avocado oil has a similar fatty acid profile to olive oil but a very low level of FFAs – typically less than 0.1% – and a correspondingly high smoke point of 270 °C/520 °F. FFA levels are kept so low in the avocado because the oil pockets in the plant cells have less contact with water than in an olive plant. Differences in ripening pathways exist for the two fruits. Avocados are described by horticulturalists as being climacteric – that is, they show a burst of respiration and ethylene production as plant starch is converted to sugar during ripening. Most climacteric fruit can ripen on or off the tree or vine, but avocados are unusual in that they ripen easily *after* harvesting. This allows farmers to pick the fruit in its more sturdy unripe state and allow ripening to happen under controlled conditions at a time convenient for marketing or pressing. Olives, by contrast, are non-climacteric and don't show such a ripening profile. They must ripen on the tree and can be easily bruised or spoiled in the harvest process.[15] Once harvested, farmers have only hours (some might say days) to get them to the press before degradation begins and FFA levels rise. These differences explain why the smoke point of avocado oil is so much higher than olive oil, despite many other similarities.

9.1.3   Olive Oil as Survival Rations

When Tim Jarvis and his partner Peter Tressider decided to be the first humans to cross the Antarctica with a dogsled in 1999, providing sufficient calories for the crossing was an issue. They estimated that each of them would need to eat approximately 7000 calories a day to fuel their survival against the cold and vigorous exercise. The answer? Olive oil. Jarvis reported that they consumed "… roughly 50 mills at breakfast on your cereal, 50 mills in the soup, and then 50 mills in the evening meal each. So it's a big glug of olive oil each day." Their successful completion of the historic trip is testimony to the power of olive oil to be the most efficient source of calories on a per gram basis![16]

We were told of a couple in their seventies who decided to not let age stop them and complete a life long dream of thru-hiking the Pacific Crest Trail from the Mexican Border northward to Manning Park, British Columbia. To keep the weight of their supplies down for this arduous 2600 mile journey that can take six months, the couple decided to forego normal camping food and, instead, take all of their caloric intake through olive oil. We are not sure how they celebrated completing the trip but my guess is they chose NOT to drizzle olive oil on their potatoes.

9.1.4   Olive Oil as a Food Preservative

Olive oil was used as a food preservative before the widespread availability of refrigeration and modern food handling techniques. Antimicrobials in the olive oil act to naturally sterilize food – one example is a study done on the death rate of *Salmonella* in mayonnaise made with various oils.[17] In the study, mayonnaise made with olive oil showed the highest auto-sterilization. It would, however, not be prudent to rely on this method to keep your potato salad from giving you food poisoning on a hot summer's day.

In some cases, olive oil was even used to help to protect fine bottles of wine on the longest journey imaginable – that to the Underworld. The preservation of the liquid remnants in the Speyer Wine bottle, the oldest unopened wine bottle in the world (vintage circa 325 AD) is attributed in part to olive oil that topped the originally bottled wine.

Figure 9.4 shows the thick layer of olive oil on the top plus the wax sealed cork that were meant to protect the wine through the difficult uncertain journey that its owners – a dead Roman nobleman and his wife – were about to take to the underworld. The bottle was found during a 19th Century excavation of a 4th Century Roman village in Germany. Scientists advise that, although the liquid contents are likely to still be microbiologically intact, the wine is not likely to be pleasing to the palate. The olive oil served as both a physical barrier to oxygen and evaporation and a chemical barrier to oxygen through the antioxidants present in the oil, a role that sulfur dioxide added at bottling serves today.[18] The bottle remains unopened and on display in the Tower Room of the Historical Museum in Speyer, Germany.

Figure 9.4   The 1700 year old Speyer Wine Bottle, the oldest unopened wine
bottle in the world, was preserved by topping the wine with a layer
of olive oil. Image from Wikimedia © CC BY-SA 3.0, photo credit
Immanuel Giel 2005.

A Japanese island called Shōdo Island, or Olive Island, has
groves of olive trees, a press, and some EVOOs that take top honors in international competition.[3] These olive trees were originally grown and the olive oil processed not for olive oil, but to
act as a preservative for the fish products that were a resource for
the islanders and a delicacy that sold well back on the mainland.
Ikura, or salmon roe, is often used in sushi and sashimi and is
classically preserved in olive oil.

9.2   IN THE DARK

It has been said that lamps are one of the very first human inven-
tions – with some archeological evidence dating back to the 12th
Millennium BCE. The first evidence of a lamp that used olive oil
as a fuel emerges from archeological sites around Israel and dates
to the Early Bronze Age, or from 3300–2200 BCE.[1] Olive oil lamps
of the type seen in Figure 9.5 were apparently very common, if

Figure 9.5 Replica of an Ancient Greek olive oil lamp by Epalladio art workshop ©2004–2012, http://www.epalladioartworkshop.com/ OILLAMPS/ used with permission.

judged by the number of them that survive the thousands of years to appear in museums around the Mediterranean. All you need is some oil and a wick, and a beautiful lamp can be made.

This use of olive oil is, of course, the origin of the name for the lowest grade of oil: lampante. We have learned from conversations with artists in Turkey that even the soot from the olive oil lamp has its uses: it makes an excellent pigment to darken oil paints and pastels.

9.3   IN THE MEDICINE CABINET

The use of olive oil as creams, unguents, balms, salves, infusions, and plasters is just as old as its use as a food. Since the tree itself was a gift from the Gods, the ancients used every part of the sacred olive tree for medicinal uses: leaves, bark, wood, fruit, seeds, as well as the oil (see Table 9.2 for some examples). Today, we know that oleuropein from the leaves, lignans from the wood, ligstroside from the bark, all display bioactivity while the fruit, seeds, and of course, the oil are also rich in medicinal compounds. These have been used to alleviate a host of ailments: olive oil has been used as an analgesic to fight pain, an anticonvulsant against seizures, an anti-inflammatory in the treatment of diabetes, an antihypertensive to lower high blood pressure, an antimicrobial to accelerate wound healing and fight infection.[19]

In the 2nd Millennium BCE, Mycenae was an important Greek city southwest of Athens in the Pelopennese Islands. A list excavated from the Mycenaean ruins contains a catalog of all of the

Table 9.2   Some of the medical uses of olive products. Copyright ©
2015 Muhammad Ali Hashmi *et al.* Creative Commons
Attribution License.

S. no.	Part/preparation used	Aliment/use
1	Leaves and fruits/infusions and macerations	Hypoglycemic, hypotensive
2	Decoction or infusion of fruits and leaves	Antidiabetic
3	Olive oil + lemon juice	To treat gallstones
4	Oil of seeds/taken orally	Laxative
5	Decoctions of dried leaves and fruit/ oral use	Diarrhea, respiratory, and urinary tract infections
6	Olive oil/applied on scalp	To prevent hair loss
7	Boiled extract of fresh leaves/taken orally	To treat asthma
8	Boiled extract of dried leaves/taken orally	To treat hypertension
9	Leaves extract in hot water	Diuretic
10	Olive oil	Applied over fractured limbs
11	Infusion of leaves/oral use	Antipyretic
12	Olive fruit	Skin cleanser
13	Infusion of leaves/oral use	Anti-inflammatory, tonic
14	Leaf preparations	To treat gout
15	Leaves of *O. europaea*	Antibacterial
16	Decoction of leaves	Antidiabetic, antihypertensive
17	Fruits and leaves	Hemorrhoids, rheumatism, and vasodilator
18	Infusions of leaves	Eye infections treatment

plants whose olive oil extracts could be used in the preparation
of healing ointments. The list included celery, fennel, water-
cress, mint, sesame, sage, rose, and juniper[20] by extracting the
olive-soluble active components into the oil, their health giving
properties would be made more readily available to humans.

Hippocrates, the Greek father of modern medicine, wrote in the
4th Century BCE of over 60 medical uses of olive oil.[21] Included
in his list were both external uses – healing of dermatological
diseases, lacerations and burns, gynecological diseases, and ear
infections –and internal uses, such as olive oil's use as an emetic
(inducement of vomiting) and to address indigestion and other
gastrointestinal issues. Aromatherapy was an important tool in
early health care, and recipes for producing heated infused oils
whose very vapors would calm and heal can be found in ancient
texts.[2] In the 1st Century AD, Roman naturalist Pliny the Elder

Figure 9.6 Olive oil medicinal label *circa* 1920 (image in public domain).

advocated the use of olive oil for medicinal purposes, but not just any olive oil, instead that "obtained from the raw olive and when it has not begun to ripen."[22] We know today that early harvested olives yield oil with the highest levels of antioxidant polyphenols.

In the early to middle 20th Century, it was not uncommon to find olive oil on the shelves of pharmacists and chemists in the USA and abroad. Its formulation was likely to be a liquid, to be taken orally by the spoonful for gastrointestinal issues or high blood pressure, or topically as needed to help alleviate muscle pain, accelerate wound healing, or relieve itching as a result of skin ailments. A typical druggists label for olive oil is shown in Figure 9.6.

Today, with growing interest in natural products, one can find many online references to using olive oil as part of an oral health regime in a practice known as olive pulling. By pulling the oil into the mouth and swishing for some period of time, toxins are said to be extracted from the mouth and presumably bacteria are killed. This practice is recommended not just for olive oil but also for several other edible oils, including sesame and coconut. To date, there is little scientific evidence that backs these claims and the American Dental Association recommends that instead of oil pulling, a regular oral health regimen should include brushing of teeth twice a day, and flossing and the use of an antiseptic mouthwash once a day.[23]

9.4   IN THE SPA

9.4.1   As a Body Cleanser and Soap

A very common way to remove sweat and dirt before soap was in common use was to apply olive oil to the body and then scrape off the suspension off dirt and oil using a slightly curved spatula known as a strigil. Depictions of this type of cleansing can be found on many ancient jugs and paintings and strigils are commonly found in tombs, where they would have been part of a "personal hygiene kit" necessary for the afterlife.

Remnants of actual soap were found in clay vessels in Babylon dating back to 2800 BCE. A recipe found in Babylonian tablets from 2200 BCE specified that soap could be made from fat, water, and wood ash. These early soaps were primarily used as general cleaners for wool and other materials, and not for bathing. A thousand years later, Egyptians were using soap for cleaning their bodies, and by 79 AD, the citizens of Pompeii had a soap factory, as can be told from the ruins of that great Roman city. As time passed, soap making became more and more refined, and by the 17th Century AD, Marseilles soap, a French formulation using olive oil, Mediterranean salt water, and lye became popular for its consistently high quality.[24]

Soap making is a quite straightforward chemical process in which an acid (the fatty acids from fat or oil) is mixed with a base (from the wood ash or some other alkaline plant). When adding a base to a triglyceride (TAG), the reaction is slightly more complicated, but is known as "saponification" (literally, soap-making). First, the fatty acid is broken away from the glycerol by the addition of the base in water.

Under these conditions, the FFA not only breaks off, but it also loses its acidic H, producing a fatty acid anion, as shown in Figure 9.7 that reacts with any positive counter ion – like the sodium (Na^+) in sodium hydroxide (NaOH), to make a salt – the soap. The chemical reaction shown in the left panel of Figure 9.8 shows that the three FFAs, originally connected to the TAG, each require one molecule of base to be added to fully react, so three equivalents of base must be added for each TAG. The TAG also produces another product – glycerin – that is also an effective skin softener. The right panel of Figure 9.8 shows the molds we used to shape the soap during drying. Wood ash was a common base used by the ancients, but by

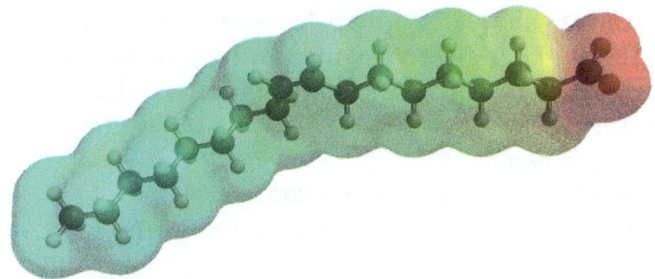

Figure 9.7 Structure of oleate, the anion of oleic acid found in soap. Note the very polar group on the right end, and the nonpolar tail on the left side. That property – polar on one side and nonpolar on another, makes the molecule soluble in both greasy stains and in water.

Figure 9.8 Chemistry of soap making.

the time soap production was established in Marseilles, France, in the 1600's, certain basic plants, such as marsh samphire and saltwort, were used. Eventually, lye (commercial NaOH) replaced the botanically derived base in modern commercial soap production.

The different steps in soap making have changed very little over time, as this recipe from a 17th Century Marseilles soap maker suggests.[25] While this process included a heating step, it is possible to make soap in a cold process with slightly less reliable results:

1. *Pasting of the oils*: emulsion of soap fats and oils with lye. The mixture is brought to the boil in enormous cauldrons.
2. *"Epinage"*: removal of impurities from the bottom of the cauldron, carried out three times.

3. *Finishing*: the lye is boiled for several hours, then impure sediment is removed again and clean water is used to rinse the remaining impure sediment to the bottom of the cauldron.
4. *Drying in molds.*
5. *Cutting while the soap is still soft.*
6. *Stamping*: manufacturer's name and brand after solidification.

Students in our classes have made a safer cold processed soap, as shown in Figure 9.8, to take home with them – scenting it with rose oil or bergamot and coloring it if desired.[26] For many immigrants from olive oil countries, the pure scent of olive oil soap brings back sweet memories from home.[27]

9.4.2 As a Skin Cream and Hair Treatment

Olive oil has been used as a skin cream and as a cosmetic for thousands of years. Pliny claimed that while wine was good for the inside, olive oil was particularly good for the outside of the body.[22] Olive oil may be used right out of the bottle, or it may be mixed with other substances such as beeswax or coconut oil. It may also be scented with rosewater or essential oils from lavender or eucalyptus. Many different recipes exist for making skin and hair care products from olive oil. A simple night cream can be prepared from the following recipe:

Scented olive oil night cream[28]

Ingredients
- 1/4 cup extra virgin olive oil
- 1 tablespoon coconut oil
- 1 teaspoon beeswax
- vitamin E (about 2 capsules)
- 2–5 drops rosewater, lavender or other essential oil (optional)

1. Combine the first three ingredients and heat on low, stirring occasionally, until everything is melted. Add the vitamin E and a few drops of essential oil.
2. Cool the mixture until solid and then whip until fluffy. An egg beater or single beater of a hand mixer can be used.

3. Store night cream at room temperature. If it gets too warm and melts, just cool it and it will solidify again. This will not affect the cream in any way. Cream will last a couple of months at room temperature due to the antifungal/antimicrobial components of the coconut oil.

Olive oil's soothing effects on the skin also make it a natural choice for addressing problems with dry hair or dry scalp. A simple drop or two of olive oil on your hands and then spread through your hair will take care of the flyaway hair that particularly occurs in dry winters. When a full spa treatment is desired, making an emulsion of olive oil with egg yolk and honey, combing this through your hair and letting it sit for an hour or two before shampooing creates a glossy luxurious head of hair. Some even believed that olive oil rubbed into the scalp could help with baldness.

9.4.3   As a Base for Perfume

Olive oil was used as a base for making perfumes because it dissolves many pleasant smelling aromatic compounds and tends to stay on the human skin for quite some time.[2] Much evidence exists to suggest that the primary use of perfumes by the ancients was to attract the attentions of an unwary mate. This feature of perfumed oil seems to transcend time and space: the Greek poet Homer describes Hera's seduction of Zeus in ancient Greece (between 12th and 8th Century BCE); Egyptian papyral writings describe Istar, Queen of Nineva's seduction of Hedammu, and the Bible describes Ruth's seduction of Boaz. Each used the anointing of scented (olive) oil both as an aphrodisiac to capture someone's attention and signal a willing and cooperative mate. Not to leave the seductive act entirely in the hands of females, legend (not history) tells of the seduction of the wife of the pharaoh by the Egyptian God Amom who, through the use of fine perfumed olive oil in great concentration, attracted the attention and finally the conjugal bliss of Egypt's noble lady. The offspring of that union was Hatshepsut, the first female pharaoh, who came to the throne in 1476 BCE.

9.5   ON THE PERSONAL SIDE

A natural concern in ancient times was how to prevent conception.[29] Hippocrates believed rubbing the uterus with old olive oil prior to intercourse would prevent conception. Aristotle noted contraception could be achieved by "anointing" the uterus with a combination of olive oil mixed with cedar oil, white lead, and frankincense (please, do not try this yourself – we know a lot more about how toxic lead is today). While Pliny held the beliefs of his time that contraception and abortion were not to be encouraged in the first case and illegal in the second, his treatise Natural History noted a contraceptive cream could be prepared from olive oil, pigeon droppings, and wine. A second method he reported, perhaps with his Roman tongue in his Roman cheek, was an amulet that was to be worn for a year that contained worms from the head of the hairy spider (XXIX, Chapter 27).[22,30] Naturally, we do not condone any of these methods but include them to provide a historical perspective.

Since olive oil consumption is associated with a healthier cardiovascular system, it should also help address symptoms of erectile dysfunction, not in the acute manner of a medication like Viagra, but as a correlate of the improvement of your general cardiovascular health. One Spanish study[31] examined the therapeutic effect of consumption of lycopene from a mixture of tomatoes and olive oil.

Vaginal dryness can be addressed with the application of as little as five drops of olive oil. The long-lasting lubrication provided by the oil is a welcome safe choice for women looking to reduce the pain of intercourse due to vaginal dryness. The documented use of olive oil as a personal lubricant for intercourse dates back, as you could guess, to the times of the Greeks and Romans. The ancients did not have to worry about the latex used in condoms, which is broken down by olive oil. We know today that these materials don't mix successfully.

9.6   ON THE ATHLETIC FIELDS

The ancient Olympic Games were originally held on the plains of Olympia, Greece, and can be accurately traced back to 776 BCE, though it is likely that they began much earlier. Coroibos, a cook from Elis, who won a foot race in one of the early Olympics, was no doubt familiar with olive oil and was pleased when several jugs of fine olive oil were awarded as a prize. At these events, winners would be given a wreath of olives to wear on their heads. The athletes themselves would often be anointed with oil before the competition even began, so as to best show off their physique no doubt, but one can't forget the fact that olive oil contains an anti-inflammatory compound, oleocanthal, that will have pre-emptively helped the Olympic athletes with their muscle pain. Images from the time show athletes using a scraper to cleanse themselves – likely after the events – as in keeping with the hygienic practices of the time. Pliny, while praising the olive and citing the high honor that the Romans have conveyed upon it, laments its use "for luxury" by the Greeks (Book XV, Chapter 5, p. 284).

"The Greeks, those parents of all vices, have abused it by making it minister to luxury, and employing it commonly in the gymnasium: indeed, it is a well-known fact that the governors of those estab-lishments have sold the scrapings of the oil used there for a sum of 80 000 sesterces."

In case you are not familiar with the exchange rate, one ses-terius is a unit of money equal to 1/4 Denarius. Eighty thousand sesterces is quite a bit of money for some bodily scrapings – it amounts to the equivalent yearly salary of almost 100 Roman legionnaires or the purchase of about 30 slave girls. Pliny goes on to document the Greek's use of this "abominable collection of filth" as follows (Book 28, Chapter 13).[22]

"... the owners of the gymnasia have introduced the very excretions even of the human body among the most efficient remedies; so much so, indeed, that the scrapings from the bodies of the athletes are looked upon as possessed of certain properties of an emollient, cal-orific, resolvent, and expletive nature, resulting from the compound of human sweat and oil. These scrapings are used, in the form of a pessary, for inflammations and contractions of the uterus: similarly employed, they act as an emmenagogue, and are useful for reducing condylomata and inflammations of the rectum, as also for assuaging pains in the sinews, sprains, and nodosities of the joints. The scrapings obtained from the baths are still more efficacious for these purposes, and hence it is that they form an ingredient in maturative prepara-tions. Such scrapings as are impregnated with wrestlers' oil, used in combination with mud, have a mollifying effect upon the joints, and are more particularly efficacious as a calorific and resolvent; but in other respects their properties are not so strongly developed."

Today, the great sport of olive oil wrestling exists much as it has for the last 1000 plus years. The sport is exactly what it sounds like. Men strip down to specialized leather leggings made from buffalo hides called a "kispet," liberally cover themselves with olive oil, and wrestle (Figure 9.9). This sport traces its roots back to the Ottoman Empire and the annual competition called Kırkpınar after the site near Edirne, Turkey where the first match was held in 1362. It is reported to be the world's oldest continuously running sporting event.[32] UNESCO includes Kırkpınar among its list of "Intangible Cultural Heritage of Humanity" describing it as "strongly rooted in the practitioner community as a symbol of identity and continuity highlighting the virtues of generosity and honesty and reinforcing members bonds with tradition and custom, thus contributing to social cohesion and harmony."

In 2016, 1969 wrestlers competed for the coveted "Golden Belt" in the competition that is part showmanship as the master of ceremonies or "Ağa" first introduces each player by singing his name, history, and background. With prayerful ceremony, the players are then liberally anointed with olive oil (2000 kg of oil was used in 2016). Wrestlers begin their event by touching the ground to honor their origins and connections to the earth and then formally greet each other with ritualized gestures. The athleticism of the competitors shows up in the 40 minutes they each have to pin their opponent's shoulders to the ground, carry him

Figure 9.9   Kırkpınar wrestling match in Turkey, the longest continuous running athletic competition in the world. Public domain; File:Yagli gures2.jpg.

above shoulder height, or rip his opponent's leather britches.[33] Four time champion Recep Kara became the 655th Kırkpınar gold belt winner in 2016, 12 years after he became the youngest winner ever at the age of 22.

9.7   IN OUR MOST HOLY PLACES

9.7.1   Ritual Anointing

In the Judeo Christian tradition, anointing with Holy Oil was a means of setting apart an object, a person, or a place from its ordinary use. God hands Moses down a prescription for Holy Oil in Exodus 30: 22–30 (New King James Version)

> *Moreover the Lord spoke to Moses, saying: "Also take for your-self quality spices—five hundred shekels of liquid myrrh, half as much sweet-smelling cinnamon (two hundred and fifty shekels), two hundred and fifty shekels of sweet-smelling cane, five hundred shekels of cassia, according to the shekel of the sanctuary, and **a hin of olive oil**. And you shall make from these a holy anointing oil, an ointment compounded according to the art of the perfumer. It shall be a holy anointing oil. With it you shall anoint the taber-nacle of meeting and the ark of the Testimony; the table and all its utensils, the lampstand and its utensils, and the altar of incense; the altar of burnt offering with all its utensils, and the laver and its base. You shall consecrate them, that they may be most holy; whatever touches them must be holy. And you shall anoint Aaron and his sons, and consecrate them, that they may minister to Me as priests.*

Even the name of the Son of God, Jesus Christ, contains the Christ – which is derived from Kristos – the anointed one. Jesus was to be seen as holy and set apart – anointed and protected by his Holy Father.

9.7.2   At Birth and in the First Few Months...

Olive oil is reported to have been used in ancient times to liber-ally lavage the birth canal during a difficult birth so as to ease the baby's passage into the world. Today, it is a common practice in some cultures to massage an infant with oil at birth and in the first few months of a child's life. This is thought to help stimulate circulation, alleviate dry skin. We were not surprised when a pediatrician in Istanbul advised the use of olive oil rather than diaper cream to prevent diaper rash for a newborn.

Pediatricians often worry about transepidermal water loss (TWL) in newborns and typically will recommend topical application of an emollient – a stabilized mixture of oil and water. These emulsions can either be creams (oils suspended in a water base) or ointments (water suspended in an oil base). The particular oil that one uses is very culturally dependent. One study in neonatal practices reported that olive oil is the most frequently used by midwives and its use is passed down from generation to generation (82%). However, these researchers reported that olive oil can be harsh on the skin barrier of babies, and instead recommend a high linoleic acid oil such as sunflower seed oil or almond oil.[34] Other studies concur with this assessment, stating that a few cases of contact dermatitis involving newborns and olive oil have been reported.[35] So please consult with your own health care providers before applying any olive oil to a young one's skin for these purposes.

9.7.3   At Death

On the ancient Egyptian sarcophagus of Ankhnsneferybra of the 26th Dynasty, Shesmu, the Demon God of Egypt, is depicted as the manufacturer of the Oil of Ra.[36] He was thought to be the Master of Perfumes because of the way the Egyptians infused oil to create their perfumes. It was in his role as a god of perfume that he was linked to the mortuary cult. Not just a god of the underworld, he was also a god who provided the sacred oils for the embalming process. It was believed that he prevented the putrefaction and decay of the flesh after death with his unguents and special oils.

When oil was not enough to protect, olive oil was used in laying out the dead. It was typically the women's job to wash the dead body and anoint it with olive oil or scented oils.

Honoring the dead, preparing them for the afterlife, and in the case of very holy or exalted persons, preparing their remains to provide solace and salvation for their followers was serious business for Greeks and Romans as well. A necropolis discovered in Sidon in 1887 with a tomb 23 centuries old contained the body of Tabnith, Priest of Ashtaroth, and King of the ancient Phoenicians[37] Opening the tomb, archeologists observed the following

> *"The royal corpse lay in a sort of liquid resembling (olive) oil. ...the body had a fresh natural appearance and was well preserved. The flesh was soft to the touch."*

The Museum of History in Marseilles has an excellent exhibit from the excavated remains of a Roman necropolis that was built adjacent to the ancient city of Marsylus – site of modern day Marseilles. As shown in Figure 9.10, the necropolis contained sophisticated sarcophagi, with several layers of internal containment leading to a final grand stone entombment. One particularly large and lavishly carved sarcophagus on display had

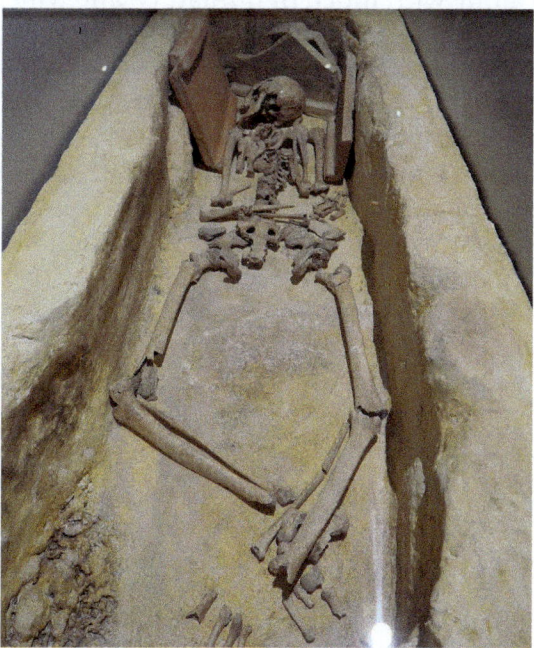

Figure 9.10   Olive oil was used in preserving the dead. Human remains were entombed in a sarcophagus and olive oil flowed through from a port in the top and out the bottom, as shown in the burial tombs in the Museum of History in Marseilles, France. © 2016 http://www.Mikestravelguide.com, used with permission.

a curious pipe-like spigot projecting from the bottom corner. A corresponding liquid inlet port could be seen on the top cover of the stone. Chemical analysis of the residue on the interior of the sarcophagus revealed the liquid to be olive oil. It is not clear whether submerging the dead in olive oil was used as one stage in preparing the body for the afterlife or whether oil poured over the bones of the revered dead, and then drained into a vial, was a source of a blessed oil to use in ritual anointing. This custom of infusing olive oil with a holy essence by passing it through a tomb containing the bones of the holy dead and collecting it from a spigot into a vial is described as an early 20th century practice in central Anatolia.[38]

9.8 SUMMARY

How can one summarize the diversity of uses of a substance that has evolved along with much of humanity, and which touches us literally from cradle to grave? While we have mentioned many uses that have fallen from favor (we note that few strigils are available at our local bath products store), we have described many that are commonly in use today. We have tried many of these, and found them effective, whether it was a family tradition or newly learned. Olive oil really can make our hair more manageable, or small cuts heal nicely. It makes a wonderful soap and, indeed, a beautiful light. All of these uses are in addition to the magical effect it has on food: making many foods both look and taste better.

With such a powerfully useful product, the only question that remains is: can we continue and spread this relationship between humans and the olive in our modern world? We will address that question in our final chapter.

REFERENCES

1. N. Liphschitz, R. Gophna, M. Hartman and G. Biger, *J. Archaeol. Sci.*, 1991, **18**, 441.
2. M. Brumer, *J. Intercult. Interdiscip. Archaeol.*, 2015, 89.
3. *Olive Oil Pairing App Finds the Best Olive Oil*, http://www.olive-oiltimes.com/olive-oil-basics/olive-oil-pairing-app-finds-the-best-olive-oil/50264, accessed July 2016.

4. R. Verkerk, M. Schreiner, A. Krumbein, E. Ciska, B. Holst, I. Rowland, R. De Schrijver, M. Hansen, C. Gerhäuser, R. Mithen and M. Dekker, *Mol. Nutr. Food Res.*, 2009, 53(suppl. 2), S219.

5. J. P. Ramírez-Anaya, C. Samaniego-Sánchez, M. C. Castañeda-Saucedo, M. Villalón-Mir and H. L. de la Serrana, *Food Chem.*, 2015, **188**, 430.

6. W. Stahl and H. Sies, *J. Nutr.*, 1992, **122**, 2161.

7. T. W.-M. Boileau, A. C. Boileau and J. W. Erdman, *Exp. Biol. Med.*, 2002, **227**, 914.

8. J. M. Fielding, K. G. Rowley, P. Cooper and K. O'Dea, *Asia Pac. J. Clin. Nutr.*, 2005, **14**, 131.

9. M. Bubonja-Sonje, J. Giacometti and M. Abram, *Food Chem.*, 2011, **127**, 1821.

10. H. Rasoolijazi, M. Mehdizadeh, M. Soleimani, F. Nikbakhte, M. Eslami Farsani and S. Ababzadeh, *Med. J. Islam. Repub. Iran.*, 2015, **29**, 187.

11. S. Bastida and F. J. Sánchez-Muniz, *Food Sci. Technol. Int.*, 2001, 7, 15.

12. Y. Allouche, A. Jiménez, J. J. Gaforio, M. Uceda and G. Beltrán, *J. Agric. Food Chem.*, 2007, **55**, 9646.

13. S. Cicerale, X. A. Conlan, N. W. Barnett, A. J. Sinclair and R. S. Keast, *J. Agric. Food Chem.*, 2009, **57**, 1326.

14. *The Science of Cooking with Olive Oil*, http://www.truthinoliveoil.com/2013/10/science-cooking-olive-oil, accessed July 2016.

15. K. Thompson, *Fruit and Vegetables: Harvesting, Handling and Storage*, John Wiley & Sons, 2008.

16. *Antarctic Crossing to Rely on Olive Oil*, http://www.abc.net.au/am/stories/s59275.htm, accessed July 2016.

17. N. J. Russell and G. W. Gould, *Food Preservatives*, Springer Science and Business Media, 2003.

18. R. S. Jackson, *Wine Science: Principles and Applications*, Academic Press, San Diego, 1994.

19. M. A. Hashmi, A. Khan, M. Hanif, U. Farooq and S. Perveen, *Evidence-Based Complementary Altern. Med.*, 2015, **2015**, 541591.

20. *Olive Oil in Ancient Greece*, http://www.greekolivespa.com/en/olive-oil-in-ancient-greece.html.

21. Hippocrates, *Hippocrates Collected Work*, Harvard University Press, 1868.

22. Pliny the Elder, *Natural History*, English trans. H. Rackham, William Heinemann LTD, London, 1962.
23. *The Practice of Oil Pulling*, http://www.ada.org/en/science-research/science-in-the-news/the-practice-of-oil-pulling, accessed July 2016.
24. H. Butler, *Poucher's Perfumes, Cosmetics and Soaps*, Springer, Netherlands, 2000.
25. *History of Soap Making*, http://www.marseille-tourisme.com/en/discover-marseille/tradition/the-soap/, accessed July 2016.
26. S. T. Mabrouk, *J. Chem. Educ.*, 2005, **82**, 1534.
27. J. Angus, *Olive Odyssey: Searching for the Secrets of the Fruit that Seduced the World*, Greystone Books Ltd, 2014.
28. *Handmade Olive Oil Night Cream*, http://www.jellibeanjournals.com/olive-oil-night-cream/, accessed June 2016.
29. B. Suitters, *Nurs. Mirror Midwives. J.*, 1969, **128**, 30.
30. J. M. Riddle, *Contraception and Abortion from the Ancient World to the Renaissance*, Harvard University Press, Cambridge, Mass, 1992.
31. M. Garrido, A. M. González-Flores, E. Prior, J. García-Parra, C. Barriga and A. B. Rodríguez Moratinos, *J. Sci. Food Agric.*, 2013, **93**, 1820–1826.
32. M. el-Fers and M. Kirbacoglu, *Kirkpinar – All About Turkish Oilwrestling*, Lulu Press, 2009.
33. *654th Kirkpinar Coming to Edirne, Turkey*, http://www.oliveoil-times.com/olive-oil-basics/654th-kirkpinar/48099, accessed July 2016.
34. L. F. Eichenfield, I. J. Frieden, A. Zaenglein, and E. Mathes, *Neonatal and Infant Dermatology*, Elsevier Health Sciences, 2014.
35. R. Sarkar, A. C. Inamadar, and A. Palit, *Advances in Pediatric Dermatology-2: Neonatal Dermatology*, 2014.
36. *Egypt: Shesmu, Demon-God of the Wine Press, Oils and Slaughterer of the Damed*, http://www.touregypt.net/featurestories/shesmu.htm, accessed July 2016.
37. E. A. Grosvenor, *Constantinople*, Bogazici Üniversitesi, Istanbul Turkey, 2014.
38. L. De Bemieres, *Birds Without Wings*, 2005.

CHAPTER 10

Sustainability

It's getting a bit late, and we've had a lovely dinner in the home or favorite restaurant of our host – an olive producer, perhaps, or a farmer, or a press operator. We may or may not have consumed a bit of local wine (because wine and olives frequently co-exist). As the mood turns quieter, we are likely to hear a simple question: how do we keep this going?

The reason for the question differs from place to place. Sometimes the concerns are economic – if the price paid for olive oil doesn't quite match its cost of production. Sometimes it's a worry about weather, either short or long-term. Olive producers are, at the core, farmers and have a farmer's eye for the weather and climate. Will the oil survive its trip to the consumer? After all, the producers have to trust the supply chain just as consumers do.

Very often, the concern is with people. Will the boom in demand continue, or will consumers be lured away for a new fad ("Chicken Fat Cures Cancer!"). Will we be able to find the workers we need for what we can pay them? Will the owner stay engaged, or will olive production be switched for something easier? Will the owners of an industry age out of activity without identifying the next generation of owners and passing along all the training and knowledge?

The Chemical Story of Olive Oil: From Grove to Table
By Richard Blatchly, Zeynep Delen Nircan and Patricia O'Hara
© Richard Blatchly, Zeynep Delen Nircan and Patricia O'Hara, 2017
Published by the Royal Society of Chemistry, www.rsc.org

Olive oil production is not the most lucrative career. As you have seen, it's not the easiest. If done right, you can earn enough. If done with passion and with neighbors, you can overcome the obstacles. We certainly saw no shortage of passion, collaboration, and knowledge during our travels.

What is the human side of sustainability? It's easy to forget in the blur of numbers or the weight of studies that, first of all, olive oil sustains us. After World War II, the government of Greece invited the Rockefeller Foundation to conduct a study about the social, economic, and health conditions of the Greek population. The expectation was that the population would be doing poorly; the ultimate goal was to make necessary improvements. An extremely detailed study made the surprising discovery that Cretan people were doing remarkably well! Much like today, they consumed freshly picked vegetables from their own gardens and olive oil produced from their own groves. This combination lowered their likelihood to suffer from heart disease and increased life expectancy (in comparison to the American population). Not only that but the Cretan population had a very high percentage of people over 90. To this day, the island's villagers are following the same lifestyle, which reduces the impact of the current economic crisis of Greece. It seems to us that this island had already discovered the key to sustainability.[1]

The olive tree sustained Van Gogh in an emotionally difficult time. In the end of 1889, while making many paintings of olive groves in Provence (an example is found in Figure 10.1), he wrote that:

"he was 'struggling to catch [the olive trees]. They are old silver, sometimes with more blue in them, sometimes greenish, bronzed, fading white above a soil which is yellow, pink, violet tinted orange...very difficult.' He found that the 'rustle of the olive grove has something very secret in it, and immensely old. It is too beautiful for us to dare to paint it or to be able to imagine it.'

In the olive trees—in the expressive power of their ancient and gnarled forms—Van Gogh found a manifestation of the spiritual force he believed resided in all of nature."[2]

How can a subject be so challenging, so difficult, so vital, and so nurturing all at once?

Figure 10.1 Olive trees with the Alpilles in the Background, 1889, Museum of Modern Art, New York by Vincent Van Gogh.

10.1 ABOUT THE MEANING OF SUSTAINABILITY

Sustainability can have a broad range of meanings such as sustainability of a friendship, of the financial situation of a business, or sustainability of the environment. The latter meaning of the word has been in use since 1972 with increasing public awareness and concern.[3] Remember the Inconvenient Truth? Its definition according to *The Oxford English Dictionary* is "the type of human activity and culture that will not lead to environmental degradation and depletion of natural resources." There is no shortage of dramatic scenarios where human kind uses up all the resources the earth has to offer, such as the popular movies *Interstellar* or *Mad Max*. The olive tree is literally one of the very first domesticated species on earth. Its companionship with humans has lasted for over 7000 years. The wild olive tree has sustained its existence at least for 60 000 years through wealth, wars, and famine. Its most significant characteristic is its resilience. Even in the most difficult conditions it will survive and still produce a healthy product. It is the one plant mentioned in the legend of Noah surviving the worst environmental disaster story ever told. So when talking about "sustainability" there may be something to learn from the olive tree.

10.2   OLIVE OIL IS MADE BY PEOPLE FOR PEOPLE

10.2.1   Sustaining the Business

After our many conversations around the world, we have learned to frame our view of sustainability in a broad way. We start with an important factor: the people involved in the process. Without these people, there would be no olive cultivation, no oil production, nothing to transport, and no one to consume it. People are the source of the cultural memories and traditions, and the innovations that refresh the production and use. People are, as well, the source of the problems of poor choices in production, or fraud in formulation, or carelessness in storage and use of the oil.

10.2.2   Sustaining the Economy

Of course, the production and marketing of olive oil is a business, and as such must make at least enough to cover costs of production. Economic sustainability relies on improving efficiencies in production, enhancing the price of the product, and perhaps expanding the market. While many who are involved in the production of olive oil are substantially vertically integrated (they control planting, growing, harvest, production, packaging, and marketing), that can only be done on either small or very large scale. The slim profit associated with the sale of a bottle of olive oil is usually shared among a fairly large number of parties.

10.2.3   Sustaining the Environment

Finally, we all realize that we live, work, and eat together on a single planet. The way we produce and consume olive oil will have an impact on the Earth, and the impact on the Earth from human activities will have an impact on the production of oil. Our understanding of this inter-relationship is still evolving, but few can argue that it doesn't exist. Sustainable production and use of olive oil requires a clear understanding of the challenges and opportunities associated with this tie between olive and earth.

10.2.4   Sustaining Knowledge and Passion

While it's encouraging to speak to so many knowledgeable and passionate producers and analysts, we should not rest on the assumption that we don't need to tend these qualities and extend them to others. We are fortunate that the expansion of olive production and consumption has encouraged many to join the community of those who love olive oil. As they join, they bring new energy and a new perspective on our practices.

Science, largely chemistry, has played a major role in laying a foundation for answering these questions. We know so much more now than a few decades ago about growing olives, and processing them into oil, thanks to the chemists' increasing ability to analyze complex mixtures. The identification of so many natural products in the olive and the oil, and the connection of these compounds to their health effects has deepened our belief that extra virgin olive oil (EVOO) is intimately linked to human health. As studies of new concepts in health continue (including the study of the gut microbiome), we fully expect that this belief will continue to deepen. Knowing that the oil is good for you, and understanding why, helps to fuel the passion.

10.2.5 Sustaining Trust

Amidst the flurry of news reports about fraud in olive oil (not a new problem by any means), consumers and producers alike are looking carefully at the practices along the supply chain. And, if they are honest, they will look at their own practices.

The increased volume of this conversation most likely reflects an increased understanding of the vital role played by EVOO in diet and health. Fortunately, careful and systematic study of the chemistry of EVOO and its application to authenticated testing can help assuage concerns.

We all have a role in sustaining trust in the process. Training consumers to taste their oil and reject flawed oil will help, as will helping restaurant owners and marketers to handle the oil properly. An increased effort to specify acceptable oil and test for it by importers, and best practice by producers and exporters will help. These are goals, which are clearly some way off. They can also be expensive to implement. However, it does seem that there is movement in the right direction, on the whole.

10.2.6   Sustaining Cultural Heritage

The importance of preserving cultural heritage is important in establishing and valuing our relationship with the past. Many of our best practices today mirror the behaviors we have seen described in 2000 year old documents. Archeologists who discovered the ancient olive oil mill in Klazomenai (Urla, Turkey) dating back to the 6th Century BCE gratefully admitted that they could never have guessed the working principles of the mill from a few bare rocks if they haven't seen similar mills still functioning in the region.

Celebrating our agricultural legacies is important in a world where an increasing number of us are city dwellers. In 2014, a group of university students in Turkey, members of Zeytince Association for Ecological Living in İzmir, traveled through 700 km of western Turkey and located about 100 monumental trees between the ages of 300–1000+ and 80 ancient mills. These data have been integrated into a greater Routes Project of İzmir Municipality and are now available as an Olive Route to help tourists visit the ancient sites and groves and gain a deeper appreciation of their heritage.[4]

The connection between inhabitants in a region and that region's olive trees is also an important part of maintaining an eye on the environment. If the olive trees are viewed as part of your family, you will naturally check in on them, celebrate when they are prosperous, and mourn when they are sick or die. Likewise, knowing about the people who tend the trees, harvest the olives, and produce the oil can lead to a protective feeling about the production of oil. It is this proprietary feeling that made the destruction of 6000 olive trees in Turkey to make way for a coal power generation plant so painful for the local inhabitants (Figure 10.2). Bulldozing centuries-old trees seemed like the deepest insult to the families that owned them. Efforts through Greenpeace and other international organizations to plant saplings to help rebuild the groves and legal efforts to compensate the small growers have helped but one wonders about the likelihood that the small grove owners can hold out against the large industrial power companies in the long run.

Another very tense case of bureaucracy threatening the economic future for olive tree owners has been developing in Italy. Last year, the EU recommended the destruction of olive trees that

Figure 10.2   Village women in Yırca, Turkey, salvage olives from trees bull-
dozed to make way for a new power plant © Umut Vedat with
Greenpeace (used with permission).

Figure 10.3   Olive grove showing *Xylella fastidiosa* infected trees in the
Solento peninsula of Italy. Efforts to contain the infection have
been extremely controversial. From R. Almeida, *Science*, 2013,
353, 346. Reprinted with permission from AAAS.

might have become infected with *Xylella fastidiosa* (Figure 10.3).
The recommendations were met with tears and anger and an
appeal was filed. This story continues to evolve as the June 2016
appeal to the EU Commission in Brussels failed and the order to
destroy the infected trees was reconfirmed. Leading Italian poli-
ticians have said that the economic repercussions will be deadly
and many cultivators will be brought to their knees.[5] One Italian
grove owner we met shrugged and said that his family would just
refuse to comply.

10.2.7   Sustaining Innovation

The archeological site at Klazomenai teaches us that using technology for pressing olives has been the norm for a few millennia at least. This technology was only replaced by centrifugation by the 1930's. It's important to recognize tradition in the production of olive oil. It is also important to recognize innovation.

Our visits to Agromillora, the test farms associated with them, and those using the innovative practice of high density farming in production were inspiring in their own way. When Saudi Arabian companies transform a desert into the world's largest olive grove with seven million trees, or olive production is practiced in the mountainous regions of Nepal, or olive production in Africa can bring jobs and better nutrition to those countries, we are seeing an industry driven not by tradition, but by innovation.

Cameo 10.1   Leandro Ravetti, Boundry Bend, Australia.

Long before we met Leandro Ravetti, Technical Director of Boundary Bend Olives in Australia, we heard of his expertise in modern high density planting and harvesting. He is valued as an international consultant and teacher of master milling courses, which, we were told, were always overenrolled. His training as an agricultural engineer and experience in both his home country of Argentina, and then Spain and, since 2001, in Australia, have given him a truly global perspective. Ravetti made it possible for us to see how Boundary Bend is vertically integrated – we toured around the labs, bottling facility, nursery, and tank farm in Lara, and then the groves, on-site lab, and press at Boundary Bend. He is shown in the image at the Murray River Irrigation Station in Boundary Bend that provides water for the 2.5 million olive trees there. Ravetti's enthusiasm for what he does and his commitment to quality inspire the people who work at Lara and Boundary Bend to do their best work. During our drive through the fields at Boundary Bend, Ravetti shared humorous stories of the kangaroos and galahs (beautiful but not very intelligent wild cockatoos) that make the vast groves their home. We felt welcome and at home ourselves. Later, discussing various details of polyphenol transformations during processing and their relationship to antioxidant effects *in vivo*, we realized this guy was the real deal.

Innovative practices require imagination, and this quality is clearly demonstrated by the success of the Zetay olive grove and mill in Aydın, Turkey. The operation is built on the skirts of the Madran mountains – at the site of a former coal mine, now planted with 13 000 olive trees – processing 150 tons of olives per year. Once the coal was depleted, the company decided to turn the land into a modern olive oil production facility (Figure 10.4). The trees are irrigated using a drip system with rainfall that collects into an artificial pond. Ecological techniques are used such as application of kaolin spray to the trees to prevent sun damage and help discourage the olive fly. Another successful experiment was using 200 free-range winged animals (ducks, chickens, geese, and turkeys) to help control mice and other pests in the orchard. The company has won an Environment Protection Award from Akdeniz University and many prestigious awards from olive

Figure 10.4   Zetay olive oil orchard, near Aydın, Turkey.

oil competitions around the world. This is a good example for the Aydın region of how land can be repurposed, of how scientists and factory owners can work together, and how wise use of resources can help produce some of the world's best olive oil.

10.2.8 Sustaining Health

Sustainability also requires health. The health of both tree and the people associated with it is an important goal. The more we discover about olive oil, the more the ancient wisdom of Pliny holds true: "But on the other hand, [nature] has not willed that we should be thus sparing of oil, and so has rendered its use common and universal by the very necessity there is of using it while fresh." Daily use of olive oil is certainly healthy, when balanced with a life similarly grounded in good practice with activity, a good diet, and reduced stress.

The challenge is, as we'll discuss below, that the value chain of olive oil production is very tight, with little room for error. It takes great attention to details and best practices to make the balance come out in your favor.

The story of the Kritsa community on Crete is inspirational in demonstrating how three pillars of sustainability – that is, people, planet, profit – are achieved in real time. Nine hundred Kritsan olive oil producers are united under a nonprofit cooperative led by Nikos Zahariadis. This cooperative has partnered up with an international for profit company called Gaea, led by Aris Kefalogiannis. Gaea markets and sells the oil internationally and invests a significant portion of the income back to the cooperative to ensure production of highest grade olive oil. The cooperative makes sure all the member producers have the best infrastructure possible to maintain healthy, bountiful trees and provides them with cutting edge equipment for harvesting and processing. During our visit, it was impressive to witness how producers harvested their trees in a few days with the help of family members and brought their olives to the factory and poured them into a giant pool of olives to be mixed with their neighbors. No one doubted that the neighbor's olives could be less good than their own! A balance was engineered into the pool to keep record of the weight of incoming olive batch, so that the producer could be paid fairly. This system, and of course the fact that the Kritsan community consumes more or less only local food, means that not only are Kritsa environs covered with beautiful greenery but also they were hardly affected by the economic crisis that hit Greece so hard in recent years.

10.3 LAWS OF NATURE

We have learned (but perhaps have not fully believed) that the use of energy is governed by a few simple rules. We know, for example, that energy cannot be created or destroyed (known as the first law of thermodynamics). We also know that the universe will always move towards increasing disorder (known as the second law of thermodynamics), which has the consequence that no process can ever run with perfect efficiency. These laws of nature can be put more prosaically as "There is no free lunch" and "You can't even break even."

A common example to explain disorder or entropy is the walls of a house falling apart. A neatly made brick wall is ordered. With time, bricks fall off randomly leading to greater disorder. An old Anatolian folk saying reminds us: "We come from dirt, we go to dirt." Thermodynamically speaking this is true. All the complex molecules that we (or the olive tree) make will eventually degrade back into the simplest and most stable molecules, such as carbon dioxide and water. Whenever humans make something by bringing together building blocks, we may feel as if we are bringing order on nature. But the cost of bringing that order is energy consumption. Larger buildings, faster cars, machinery for larger amounts of high-quality olive oil all have the inevitable consequence of increasing disorder in the universe, possibly leading to undesired consequences such as pollution of air, water, and even climate change.

We are human; inventing and making things is in our nature, yet there are so many of us (some say more than the earth can handle). With such a large population, we are destined to make changes that are potentially endangering to all species including our own. The least we can do is to make sure we use our resources wisely.

10.3.1 Carbon Footprint of Olive Oil

How can we know though whether the work done is worth the energy consumed? Money earned is certainly not the only measure. Nature is indifferent to the money we make with a certain amount of energy. In recent years, we have developed other indicators to monitor our impact on nature, such as ecological, carbon, or water footprints. These measure current environmental impact and can be used to measure progress towards sustainable operations. A carbon footprint, for example, measures greenhouse gases in terms of the area of land needed to sequester carbon.[3] It is possible to produce olive oil in a carbon-neutral way (that is, the consumption of greenhouse gasses equals their production). It may seem obvious that this is the case. After all, are we not growing trees and producing fruit, all of which fix carbon from the atmosphere? The trouble is that, while the growth of trees and fruit is undeniably favorable, the production of fertilizers, the energy used to harvest olives and process the oil, and the energy needed to transport the oil to the final market all release carbon back to the atmosphere.

Led by the Greek olive products company Gaea, several producers have moved to a carbon-neutral production strategy.[6] This requires a clear examination of the entire chain of production, from beginning to end. Growing olives, with the energy required to pump water to the trees, and to produce fertilizer and pesticides, make a surprisingly large part of the carbon footprint. Many carbon footprint calculators are available online; there is one specific for Olive Oil Producers from the International Olive Council (IOC) available online.[7] According to the June 2016 newsletter of the IOC, there is now "evidence that when the appropriate agricultural practices are adopted, the carbon sink effect (or carbon sequestration) from olive trees in the biomass and soil is much higher than the greenhouse gas emissions for the production of one unit (one liter of virgin olive oil or extra virgin olive oil)."[8]

10.4   LIFE CYCLE ANALYSIS

We can speculate and discuss all of the aspects of sustainability in general terms, but the only real answer comes from the people who sharpen their pencils (or recharge their cordless mice) and carefully account for the inputs and outputs of the process of producing extra virgin olive oil. This careful accounting is represented by life cycle analysis (LCA). LCA has been developed as a tool to measure the ecological footprint of a product from cradle to grave. The elements of this tool are represented in Figure 10.5.

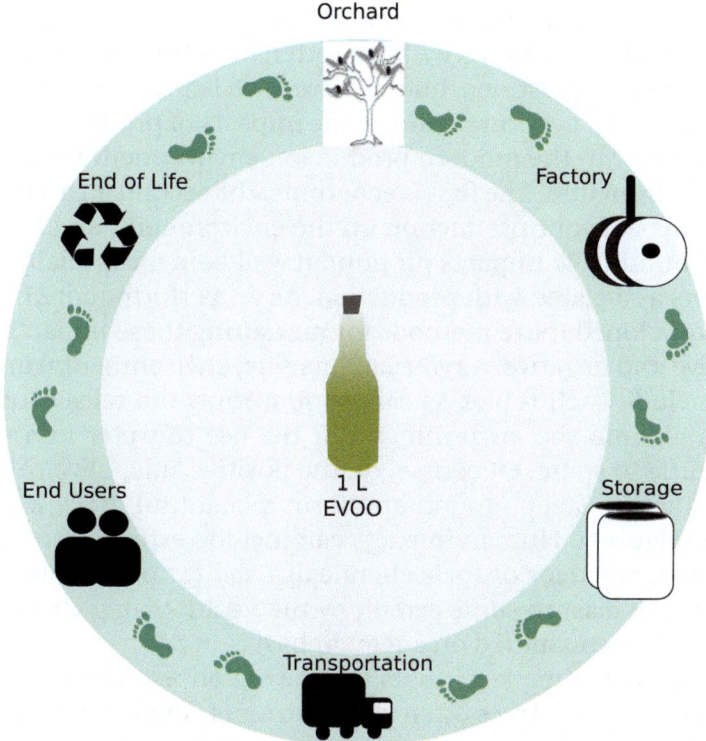

Figure 10.5 All of the considerations that must be taken into account when considering the total carbon footprint of producing extra virgin olive oil.

As soon as one begins that process, some important questions arise, as they should. First of all, we have to specify what we are producing. Of the many answers to this question, a very common choice is a one liter bottle of EVOO. More relevant to consumers than a 1000 L container, or a tiny sample, it's easy to work with. Secondly, what aspects of production should we include? Do we start with seedlings at the nursery, or with mature trees? Do we stop our analysis at the back of the truck picking up bottles from the bottling line, or try to follow the bottle into the hands of the consumer? There is no real definitive answer to these questions of boundaries, but they should be clearly defined for the analysis.

Once the object and boundaries are selected, what questions should we pursue? Economic questions are relatively easy to ask and answer: the balance in the checkbook at the end of the year tells the story fairly clearly. As we have gotten more sophisticated about these types of questions, however, we have learned that economic analysis doesn't capture many of the impacts of production on the world. Many in the modern production environment talk about a triple bottom line. The first is economic, the second an accounting of impacts of your production on the environment, and the third an accounting of impacts on human well-being, especially of the workers associated with production. As we perform more LCA, we have developed more methods for measuring these impacts, both positive and negative. As you can imagine, environmental impacts can include such topics as carbon footprint, the release of toxic materials into the environment, or the use of water in a region with little to spare. Of course, on the positive side, olive trees can help stabilize steep ground and form a beautiful grove wherever they are grown. Human impacts can include exposure to dangerous work practices or toxic chemicals used for pest control. This is offset, at least to some extent, by the health-giving character of EVOO when consumed on a regular basis.

Remember some key inputs to growing olives. Along with sunlight and rain, the trees often need nutrients, additional water, and pesticides. Making and delivering these requires energy for pumps and tractors. Harvest, similarly, requires energy for machines and trucks to transport the olives to the press. Production, though rapid, is quite energy intense and uses expensive machinery. Packaging seems to be an easy step, but where do the containers come from? They must be produced using raw materials and energy (although recycling can help offset some of this impact).

The first principle in LCA is that we need to capture everything that's done to produce the bottle of oil. Pruning with a chainsaw? What about the fuel and oil used for the activity? If adding nutrients to the tree, make sure the energy to deliver them is accounted for. The second principle is to ensure that you account for the hidden activities. Adding nutrients is a good example. It's easy to calculate how much energy is used by the truck delivering the nitrogen supplements to the farm from the store, but where did the supplements come from? The industrial process for making fertilizer can be surprisingly energy-intense. Such energy-intense activity can have a major impact on such measures as carbon footprint.[9] It is also important to find a balance between the desire to personally benefit from scientifically proven health benefits of a product that is grown half way around the world and the environmental impact of bringing it to your local supermarket from halfway around the world. Imagine the fossil fuels consumed and carbon emissions resulting from shipping those oils around the globe! No wonder the United Nations encourages investing in alternative local energy resources. Fortunately, with expansion of the areas in which the olive is cultivated, we can hope to minimize these transportation costs.

10.4.1 The Impact of Global Warming on Olive Production

What will be the impact of global warming and climate change on olive production? Scientists hasten to say that the issue is extremely complex and that it depends on many factors, some of which are difficult to model. Still, at least one study predicts that global warming may in fact lead to an increase in global olive production as increased temperatures mean increased growing days for plants and higher production – especially for a tree that is as drought tolerant as the olive tree. Some of the worst pests that lower olive oil production, such as the olive fly, are less tolerant of higher temperatures and it is likely that their population will decrease. Higher yield at a lower cost is good news for olive farmers in the long run.[10] The value of keeping the Mediterranean crops in cultivation and the biodiversity that this ensures is seen as an agricultural heritage necessary to the well-being of the planet.[11] Such a global necessity has been recognized by the establishment of the Globally Important Agricultural Heritage System (GIAHS) whose vision statement says "Dynamic conservation of all agricultural heritage systems and their multitude of goods and services, for food and livelihoods security, now and for future generations."[12]

10.5 A FINAL WORD

We believe that the olive oil industry gives back to the world as much as or more than it costs to produce a fine EVOO. We have shown how the oil can keep people alive on long expeditions, preserve ancient wines, create wonderful soaps, and just make our dinner salad taste wonderful. Trace components in the oil

Figure 10.6   Wizened old man olive tree in Turkey.

have been shown to have a host of health benefits and help add quality to a longer life. We have also seen how easy it is to destroy a good oil, either in a million dollar processing plant, in a tanker shipping the oil around the globe, or on a supermarket shelf exposed to light and heat. But when fresh and produced with care, it is delightful and has a lot of flavor. Olive oil is made by people who are interesting, engaging, and passionate, in places as widely varied as they are. This tree (Figure 10.6), the gift of the immortal goddess Athena, takes care of us in so many ways. We can pledge to do our part and take care of it as well.

REFERENCES

1. M. Nestle, *Am. J. Clin. Nutr.*, 1995, **61**, 1313S.
2. V. Van Gogh, *The Olive Orchard*, http://www.nga.gov/collection/gallery/gg84/gg84-46627.html, accessed July 2016.
3. M. Robertson, *Sustainability Principles and Practice*, Routledge, 2014.
4. *Zeytin Yolu (Olive Route)*, http://rota.yarimadaizmir.com/tr/Rotalar/2/2, accessed July 2016.

5. *Europe Orders Italians to Cut Down Olive Trees Infected with Bacteria*, http://www.telegraph.co.uk/news/2016/06/09/europe-orders-italians-to-cut-down-olive-trees-infected-with-bac/, accessed August 2016.

6. A. Kefalogiannis, *World Rev. Nutr. Diet.*, 2011, **102**, 221.

7. *Carbon Balance in Olive Oil APP*, http://carbonbalance.internationaloliveoil.org, accessed August 2016.

8. *Olive Oil Production Gives Back to Environment More than it Takes*, http://www.oliveoiltimes.com/olive-oil-basics/olive-oil-production-gives-environment-takes/52042, accessed July 2016.

9. M. Avraamides and D. Fatta, *J. Cleaner Prod.*, 2008, **16**, 809.

10. L. Ponti, A. P. Gutierrez, P. M. Ruti and A. DellAquila, *Proc. Natl. Acad. Sci.*, 2014, **111**, 5598.

11. L. Ponti, A. P. Gutierrez and M. A. Altieri, *Biocultural Diversity in Europe*, Springer, 2016.

12. *Globally Important Agricultural Heritage System*, http://www.fao.org/giahs/en/, accessed January 2016.

Subject Index

Locators in *italic* refer to figures
Locators in **bold** refer to tables